Analytical Chemistry of *Bacillus thuringiensis*

ACS SYMPOSIUM SERIES 432

Analytical Chemistry of *Bacillus thuringiensis*

Leslie A. Hickle, EDITOR
Mycogen Corporation

William L. Fitch, EDITOR
Sandoz Crop Protection Corporation

Developed from a symposium sponsored
by the Division of Agrochemicals
at the 198th National Meeting
of the American Chemical Society,
Miami Beach, Florida,
September 10–15, 1989

American Chemical Society, Washington, DC 1990

Library of Congress Cataloging-in-Publication Data

Analytical chemistry of Bacillus thuringiensis
Leslie A. Hickle, editor, William L. Fitch, editor

p. cm.—(ACS symposium series, ISSN 0097-6156; 432).

"Developed from a symposium sponsored by the Division of Agrochemicals at the 198th National Meeting of the American Chemical Society, Miami Beach, Florida, September 10–15, 1989."

Includes bibliographical references and indexes.

ISBN 0-8412-1815-3

1. Bacterial toxins—Analysis—Congresses. 2. Bacillus thuringiensis—Congresses. 3. Chemistry, Analytic—Technique—congresses.

I. Hickle, Leslie A., 1952– . II. Fitch, William L., 1948– III. American Chemical Society. Division of Agrochemicals. IV. American Chemical Society. Meeting (198th: 1989: Miami Beach, Fla.) V. Series.

QP632.B3A53 1990
632'.951—dc20 90-38800
CIP

The paper used in this publication meets the minimum requirements of American National Standard for Information Sciences—Permanence of Paper for Printed Library Materials, ANSI Z39.48-1984. ∞

PRINTED IN THE UNITED STATES OF AMERICA

ACS Symposium Series

M. Joan Comstock, *Series Editor*

Foreword

The ACS SYMPOSIUM SERIES was founded in 1974 to provide a medium for publishing symposia quickly in book form. The format of the Series parallels that of the continuing ADVANCES IN CHEMISTRY SERIES except that, in order to save time, the papers are not typeset but are reproduced as they are submitted by the authors in camera-ready form. Papers are reviewed under the supervision of the Editors with the assistance of the Series Advisory Board and are selected to maintain the integrity of the symposia; however, verbatim reproductions of previously published papers are not accepted. Both reviews and reports of research are acceptable, because symposia may embrace both types of presentation.

Contents

INDEXES

Preface

OUR FRUSTRATION WITH DISCOVERING AND UNDERSTANDING exactly what analytical methods were available and being used for *Bacillus thuringiensis* (Bt) product analysis led to our organizing the symposium upon which this book is largely based. There was no existing central forum that drew together from the various disciplines Bt researchers whose work necessitates quantifying the delta-endotoxins and exotoxins derived from the microorganism. Additionally, the Environmental Protection Agency's new Bt registration requirements identified the need for analytical assays to determine these specific components.

With the symposium, and now this book, the idea of bringing together such a diverse group of people with expertise in analyzing various aspects of Bt products blossomed into reality. Participants from the EPA, industry, academia, and government, and the fields of molecular biology, analytical chemistry, biochemistry, entomology, microbiology, and plant science were drawn together to present and discuss their findings. We would like to thank all of the authors and symposium presenters for their diligent efforts in promoting and contributing to this exciting area of biorational biopesticide biotechnology.

LESLIE A. HICKLE
Mycogen Corporation
San Diego, CA 92121

WILLIAM L. FITCH
Sandoz Crop Protection Corporation
Palo Alto, CA 94304

April 9, 1990

Chapter 1

Analytical Chemistry of *Bacillus thuringiensis*

An Overview

Leslie A. Hickle[1] and William L. Fitch[2]

[1]Mycogen Corporation, 5451 Oberlin Drive, San Diego, CA 92121
[2]Sandoz Crop Protection Corporation, 975 California Avenue, Palo Alto, CA 94304

This article presents a review of the analytical chemistry of Bacillus thuringiensis and an introduction to the remaining chapters in this book. Included are critical evaluations of the appropriate analytical targets, sample preparation and determination by bioassay, immunoassay, electrophoresis and chromatography. Issues of purifying reference standards are also discussed.

The bacterial species Bacillus thuringiensis (BT) consists of a variety of strains which are toxic to insects. Interest in products derived from these strains has undergone a renaissance in recent years due to environmental concerns associated with the use of synthetic pesticides. This has given rise to active BT research and development programs in industry, academia and government worldwide (1,2). Although there are recent reviews available on the molecular biology (3,4) and mode of action (5) of these materials, the information available on actually quantifying the active components is scattered throughout the literature. This review covers those aspects of BT technology relevant to quantifying the active product constituents in novel strains, genetically engineered plants and microbes, and in the environment.

Analytes

Rational analysis requires a thorough knowledge of the object of the analysis - the analyte or analytes. Definition of the active constituents of BT is complex because the fermentation broths which form the basis of experimental and production materials are mixtures of soluble and insoluble proteins, spores, bacterial metabolites, and growth media components. These multiple "analytes" interact synergistically to different degrees as toxins to different target insects.

The main toxins of BT are a series of structurally related proteins which are present in the sporulated cultures as crystalline inclusion bodies. These

0097–6156/90/0432–0001$06.00/0

proteins are called delta-endotoxins or insecticidal crystal proteins (icp). Different strains of the organism produce different proteins, typically multiple proteins, of molecular weights ranging from 27-140 kDa. Many of these proteins have now been sequenced and a classification scheme has been proposed which emphasizes the relatedness of the proteins, both in structural terms and in activity spectra (3). Many strains of BT, in particular the important commercial subspecies kurstaki and israelensis, contain more than one protein which may exist as cocrystals or as independent crystalline forms (6).

Determining the structures of the individual proteins in a given strain of BT is a daunting task. No chromatographic technique exists which has sufficient resolving power to distinguish between the closely related subspecies kurstaki proteins, for example. Approaches to this task include enzymatic or chemical cleavage of the mixed proteins followed by peptide mapping (see Yamamoto, Chapter 6 and Masson et al., Chapter 7), an immunological method which uses monoclonal antibodies as sequence probes (7) or an oligonucleotide probe technique which detects the presence of specific gene sequences rather than the protein itself (see Masson et al., Chapter 7).

The BT crystals are insoluble in water. They dissolve under the alkaline conditions of the insect gut to release proteins which may be proteolytically processed to yield the active toxins (8). Many of the larger protein crystals are held together by intermolecular disulfide bonds which require very high pH or reducing conditions to cleave (9). The smaller proteins, such as those of subspecies tenebrionus or san diego, are held together by hydrophobic and electrostatic interactions which can be overcome in concentrated salt solutions (10). All of the active protein toxins of BT act by binding to the insect's midgut epithelium and disrupting the integrity of membranes. The insects stop feeding and die through starvation, indirect effects of the toxin on the metabolism of the insect cells, through septicemia after invasion of the host by the spores of BT, or (in the case of subspecies israelensis) possibly involving direct muscle and neurotoxic effects (11).

Compared to other large proteins, the BT proteins in both their crystalline and soluble forms are remarkably stable to proteolytic attack and to denaturation (12). However, fermentation broths will inevitably contain variable amounts of denatured and cleaved proteins. The analyst must remember that the goal of his method is the biologically active form of the BT protein.

The spores of BT are composed of genetic material surrounded by a protein coat. At least some of the spore coats of BT strains are composed of the same protein subunits as constitute the crystals. The relative importance of crystals and spores to the potency of BT strains is very target dependent and a subject of continuing controversy in the literature (13). What can be said is that for some strains, optimum activity against some target insects requires the presence of viable spores (see also Chapter 4).

A third significant toxin of BT is the ß-exotoxin, or thuringiensin, a relatively low molecular weight metabolite produced by subspecies thuringiensis and others. This nucleotide analog is known to exert its toxic action through inhibition of DNA-dependent RNA polymerase and to act synergistically with the protein toxins (14). The determination of thuringiensin is routine by HPLC (see Chapter 13).

Finally, over the years a variety of minor, poorly characterized toxins have been described from BT species (15). These activities will not be further discussed in this chapter, but Barry Levinson in Chapter 13 describes the discovery of a novel thuringiensin analog in a subspecies morrisoni strain of BT.

Several workers have described the effects of formulation additives and small molecules on the potency of BT preparations (16,17). Particle size is known to play a significant role in the potency of BT products used against filter-feeding insects such as mosquitoes (18). There is some evidence that crystal size may be important in lepidopteran activity as well (19), possibly due to size effects on solubilization rate.

Historical Approaches to Analysis of BT

The first approach to assess the potency of BT was to count the viable spores. Based on the tendency of most strains to produce one crystal and one spore, spore count could be related to crystal count and thence to potency. This concept failed because the production of crystal protein is very media dependent. A single cell will produce crystals of varying size and varying relative proportions of its constituent protoxins based on subtle environmental changes. Spore counts are also irrelevant in assessing nonspore forming microbes or genetically-engineered plants.

The success of BT as a product of commerce has been possible due to the development of reliable laboratory whole insect feeding assays. Insect bioassay is the standard method for comparing the efficacy of different BT strains. Chemical assays can never fully match the bioassay advantage of sensitivity to all of the toxic fermentation products, as well as discrimination against denatured forms of the toxins. These methods, their strengths and their weaknesses, are the subject of Chapter 3 by Clayton Beegle.

As new chemical assay methods become available, they will inevitably be compared to the results obtained with insect bioassay. When making this comparison, it is important to realize that the goal of BT commercialization is field efficacy not lab efficacy. The correlation of field efficacy to lab bioassay is not simple. In the field, feeding inhibition, time of kill, and activity against a spectrum of organisms complicate the situation. Beegle et al. (20) have shown that laboratory potencies cannot be blindly extrapolated to the field.

A new facet to the BT business was introduced with the issuance of the EPA reregistration standard for biological insecticides in March, 1989 (see Chapter 2). The EPA is not interested in analysis of BT as a way of comparing strains or assuring efficacy. Rather their focus is on methods for enforcing that commercial products are properly labeled and safe. Bioassay is not practical for state and regional EPA laboratories. They have demanded chemical assay methods for all registered biological insecticides.

Sample Preparation for In Vitro Assays

In vitro assays are any method of analyzing BT which does not involve feeding the product to whole insects. When fed to insects, the insect gut solubilizes and proteolyses the crystals, freeing the active toxin molecules. In in vitro assays, it is necessary to presolubilize the crystals so as to present the protein in a soluble form to the determinative step, be it chromatography, electrophoresis, immunoassay, or cell culture based assay. The techniques for this solubilization are critical for assay success (9).

Solubilization of BT crystals composed of higher molecular weight proteins requires two separate processes-breaking of the intermolecular disulfide bonds and dissociation of the electrostatic and hydrophobic interactions between the protein chains. The former process requires a reducing agent such as mercaptoethanol, dithiothreitol, sulfite, cyanide, alkylphosphine or strong alkali (21,22). The latter process can be accomplished with salt solutions, high pH, low pH, urea or surfactants such as sodium dodecyl sulfate (SDS) (15,22). As mentioned previously, the BT subspecies tenebrionus or san diego protoxins lack the disulfide bonds and can be solubilized with 3 M sodium bromide solution without reductant (10).

The accepted mildest conditions for solubilization are pH 9.5 in the presence of 10 mM mercaptoethanol. However, Nickerson et al. have shown that a higher yield of protein is obtained with stronger alkaline conditions (7). Urea or guanidine yield less acceptable results and acidic solubilization has not been widely tested. Gel-electrophoresis buffer (Laemmle buffer) containing surfactant (SDS) and reductant (mercaptoethanol) is an efficient solubilizing medium for BT crystals. All of the highly alkaline solubilizing media suffer in that they promote proteolytic hydrolysis of the proteins. BT protoxins are readily hydrolyzed to the active toxin; trypsin effects this cleavage quantitatively at pH 9 (23). BT fermentations grown on complex insoluble protein matrices contain very high levels of proteolytic activity. So if the object of solubilization is intact protoxin, precautions must be taken to inhibit proteolysis. This has been accomplished with 10 mM EDTA (24), very strong alkali (see Chapters 9 and 10) or serine protease inhibitors (8).

If the object of the solubilization is the proteolytically processed toxin, then steps must be taken to assure the quantitative nature of the proteolysis. Trypsin at pH 9 will slowly dissolve and cleave intact crystals, but the process yields variable amounts of protein (25). More successful is an initial high pH solubilization to yield the protoxin followed by enzymatic cleavage. Such procedures are necessary for activation of the toxin for assay in whole cell or membrane disruption systems.

Analytical Reference Standards

The BT products of commerce are based on crude fermentation broths. They contain, at most, 10-20% of their dry weight as crystalline protein. The bioassays, which have been used until now for product assessment, have utilized powders of commercial strains as reference standards. Results of these assays are given in biological international units relative to an arbitrary value for the reference standard. Chemical assay techniques to measure the percent protein in a sample require a pure sample of the protein for reference.

Only a relatively few BT strains produce a single protein. Kurstaki strain HD73 produces a single CryIA(c) protein. Tenebrionus and san diego subspecies produce a single CryIIIA protein. Preparation of homogeneous samples of the other crystalline proteins requires genetic engineering or at least plasmid curing to yield an organism which produces only the desired protein. For this reason, mixed crystals will often be the best available reference standard.

Techniques for crystal purification are well developed (9). Bulk centrifugation separates salts and soluble metabolites from spores, crystals, and cell debris. Gradient centrifugation will readily separate crystals from heavier spores and lighter cell debris (9,26-28). NaBr gradients have the added advantage of removing proteases from the crystals (29). Dialysis or thorough washing of the crystal fractions from such preparations then yields the pure crystals. Crystal preparations should be tested to assure the complete removal of density gradient reagent. Among the BT crystals, only the CryIII protein can be recrystallized (27). The disulfide linked proteins of large crystals have never been induced to reform crystals. Other crystal separation techniques which have been found useful by various authors include two-phase partition (30-32) and flotation (33,34). To our knowledge, particle separation techniques such as sedimentation field flow fractionation (35) have not been tested on BT.

Isolated Insect Cell Line and Artificial Cell Based Assays

Activated BT endotoxin has been shown to disrupt the integrity of insect cells in culture and of a variety of artificial membrane systems such as liposomes. These effects can be detected by measurement of ion transport, conductivity, radio-, spin- or fluorescent-labeled metabolite efflux, cell death or cell lysis. Except for one brief description (1), these *in vitro* assays have not been investigated for their applicability to BT quantitation. *In vitro* assays could provide simple, quick, quantitative techniques for measuring the concentration of active toxin in a sample. Such assays could offer the specificity of whole insect bioassay at greatly reduced cost and time. They are further discussed in Chapter 5.

Immunological Methods

The first immunoassay application to BT was the rocket-immunoelectrophoresis assay (36). This technique has been largely supplanted by the ELISA techniques using monoclonal and specific polyclonal antibodies as described in Chapter 7. Radioimmunoassay of BT has been described (37). Immunoassay optimized for sensitivity may be the method of choice for detection of BT in the environment, especially for nonspore forming microbes or plants (38).

Immunoassays have long lead times for method development. Each BT strain requires the development of unique antibody reagents. These reagents have limited shelf life. Efficient ELISA methods are best suited to a manufacturing or product development laboratory with a high sample load. An ELISA is not the method of choice for an EPA enforcement lab conducting occasional assays on a wide variety of different products.

Electrophoretic Methods

The application of SDS-PAGE to BT assay is described in Chapter 9. This method is the current workhorse of analysis in this field. The methods are sturdy and reasonably quantitative. A single method can be used for all of the different strains. Their biggest drawback is in sample throughout. The methods are not amenable to automation at the current state-of-the-art.

Chromatographic Methods

The application of HPLC to the quantitation of BT subspecies san diego endotoxin is described in Chapter 8. Attempts to analyze the larger BT proteins with HPLC have not been successful due to these high molecular weight protein's tendency toward self-adsorption yielding multiple peaks.

Pyrolysis gas chromatography of purified crystals has been explored as a method for identifying and quantifying BT proteins (39).

Determination of BT in Plants

The measurement of BT genetic expression in plants is discussed in Chapter 12. Because of the low levels of expression, special high sensitivity bioassays have been developed for measurement of BT protein content in planta.

Trace Analysis of BT

As BT has been exempt from food tolerances and the attendant residue analyses, the difficult issue of trace analysis has not been important until now. Those techniques which have been used have been microbiological (spore counting, effects of specific bacteriophages on soil organisms, detection of antibiotic resistant marker genes in field trials of recombinant bacteria (40,41)). If, in the future, it is necessary to assay for BT proteins in soil and food without recourse to growth or DNA based assays, new methods, possibly sensitive radio- or fluorescence-enhanced enzyme immunoassays will be required.

Particle Analysis Based Methods

Particle or cell analysis of BT cultures has not been well studied (42). Chapter 11 describes one attempt to develop such methods.

Conclusion

As we enter the era of genetically engineered microbial insecticides, better tools for measuring the identity and quantity of individual toxic proteins are required. It is hoped that the papers in this book have presented the issues and the current state-of-the-art so as to focus the activities of the BT researcher and inspire other analytical chemists to tackle these fascinating problems.

Literature Cited

1. Jutsum, A. R.; Poole, N. J.; Powell, K. A.; Bernier, R. L. In Progress and Prospects in Insect Control; BCPC Monograph No. 43; British Crop Protection Council: Farnham, UK, 1989; p. 131.
2. Shieh, T. R. In The Impact of Chemistry on Biotechnology; Phillips, M., Shoemaker, S. P., Middlekauff, R. D. and Ottenbrite, R. M., Eds.; American Chemical Society: Washington, DC, 1988, p. 207.
3. Höfte, H.; Whiteley, H. R. Microbiol. Rev. 1989, 53, 242.
4. Brousseau, R.; Masson, L. Biotechnol. Adv. 1988, 6, 697.

5. Johnson, D. E. In Biotechnology in Invertebrate Pathology and Cell Culture; Maramorosch, K., Ed.; Academic Press: New York, 1987; p. 45.
6. Federici, B. A.; Ibarra, J. E.; Padua, L. E.; Galjart, N. S.; Sivasubramanian, N. In Biotechnology in Invertebrate Pathology and Cell Culture; Maramorosch, K., Ed.; Academic Press: New York, 1987; p. 115.
7. Höfte, H.; van Rie, J.; Jansens, S.; van Houtven, A.; Vanderbruggen, H.; Vaeck, M. Appl. Environ. Microbiol. 1988, 54, 2010.
8. Carroll, J.; Li, J.; Ellar, D. S. Biochem. J. 1989, 261, 99.
9. Nickerson, K. W. Biotechnol. Bioeng. 1980, 22, 1305.
10. Bernhard, K. FEMS Microbiol. Lett. 1986, 33, 261.
11. Singh, G.J.P.; Schouest, Jr., L. P.; Gill, S. S. Pestic. Biochem. Physiol. 1986, 26, 47.
12. Pfannenstiel, M. A.; Couche, G. A.; Muthukumar, G.; Nickerson, K. W. Appl. Environ. Microbiol. 1985, 50, 1196.
13. Li, R. S.; Jarrett, P.; Burges, H. D. J. Invertebr. Pathol. 1987 50, 277.
14. Mueller, M. D.; Harper, J. D. J. Invertebr. Pathol. 1987, 50, 201.
15. Faust, R. M.; Bulla, Jr., L. A. In Microbial and Viral Pesticides; Kurstak, E., Ed.; Marcel Dekker, Inc.: New York, 1982, p. 75.
16. Salama, H. S.; Foda, M. S.; Sharaby, A. J. Appl. Entomol. 1986, 101, 304.
17. Drobniewski, F. A.; Knowles, B. H.; Ellar, D. J. Curr. Microbiol. 1987, 15, 295.
18. Davidson, E. W. In Microbial and Viral Pesticides, Kurstak, E., Ed.; Marcel Dekkar, Inc.: New York, 1982, p. 289.
19. Scherrer, P.; Lüthy, P.; Trumpi, B. Appl. Microbiol. 1973, 25, 644.
20. Beegle, C. C.; Dulmage, H. T.; Wolfenbarger, D. A. J. Invertebr. Pathol. 1982, 39, 138.
21. Couche, G. A.; Pfannenstiel, M. A.; Nickerson, K. W. J. Bacteriol. 1987, 169, 3281.
22. Huber, H. E.; Lüthy, P.; Ebersold, H. R.; Cordier, J. L. Arch. Microbiol. 1981, 129, 14.
23. Andrews, Jr., R. E.; Bibilos, M. M.; Bulla, Jr., L. A. Appl. Environ. Microbiol. 1985, 50, 737.
24. Pfannenstiel, M. A.; Ross, E. J.; Kramer, V. C.; Nickerson, K. W. FEMS Microbiol. Lett. 1984, 21, 39.
25. Aronson, J. N.; Arvidson, H. C. Appl. Environ. Microbiol. 1987, 53, 416.
26. Mahillon, J.; Delcour, J. J. Microbiol. Methods 1984, 3, 69.
27. Li, J.; Henderson, R.; Carroll, J.; Ellar, D. J. Mol. Biol. 1988, 199, 543.
28. Sheng Zhu, Y.; Brookes, A.; Carlsen, K.; Filner, P. Appl. Environ. Microbiol. 1989, 55, 1279.
29. Nickerson, K. W.; Swanson, J. D. Eur. J. Appl. Microbiol. Biotechnol. 1981, 13, 213.
30. Goodman, N. S.; Gottfried, R. J.; Rogoff, M. H. J. Bacteriol. 1967, 94, 485.
31. Nishiitsutsuji-Uwo, J.; Endo, Y.; Himeno, M. J. Invertebr. Pathol. 1979, 34, 267.

32. Pearson, D.; Ward, O. P. Biotechnol. Lett. 1987, 9, 771.
33. Sharpe, E. S.; Herman, A. J.; Toolan, S. C. J. Invertebr. Pathol. 1982, 34, 315.
34. Yamamoto, T. J. Gen. Microbiol. 1983, 129, 2595.
35. Giddings, J. C. Chem. Eng. News, October 10, 1988, p. 34.
36. Winkler, V. W., Hansen, G. D.; Yoder, J. M. J. Invertebr. Pathol. 1971, 18, 378.
37. Tyski, S. Toxicon 1989, 27, 947.
38. Cheung, P.Y.K.; Hammock, B. D. In Biotechnology for Crop Protection; Hedin, P. A.; Menn, J. J. and Hollingsworth, R. M., Eds.; ACS Symposium Series No. 379; American Chemical Society: Washington, DC, 1988, p. 359.
39. Zhu, X.,; Li, R. Weishengwuxue Tongbao 1988, 15, 200.
40. Kearney, P. C.; Tiedje, J. M. In Biotechnology for Crop Protection; Hedin, P. A.; Menn, J. J.; Hollingsworth, R. M., Eds.; ACS Symposium Series No. 379; American Chemical Society: Washington, DC, 1988, p. 352.
41. Saldana-Acosta, J. M.; Ortega-Mendez, J. P. Publ. Biol. Fac. Cienc. Biol. Univ. Auton. Nueveo Leon 1989, 3, 15.
42. Andreev, S. N.; Brezgunov, V. N.; Bunin, V. D.; Kamenskov, B. P.; Koleznev, A. S.; Sigaev, Yu. F. Biotekhnologiya 1989, 5, 214.

RECEIVED March 26, 1990

Chapter 2

Historical Aspects of the Quantification of the Active Ingredient Percentage for *Bacillus thuringiensis* Products

George Tompkins, Reto Engler, Michael Mendelsohn, and Phillip Hutton

Office of Pesticide Programs, U.S. Environmental Protection Agency, 401 M Street SW, Washington, DC 20460

The history of regulation as it applies to the active ingredient content of Bacillus thuringiensis insecticides is discussed.

Bacillus thuringiensis (Berliner) is a gram positive aerobic soil bacterium characterized by its ability to produce crystalline inclusions during sporulation. Although this bacterium is known to have four kinds of toxin (1): the alpha-exotoxin, beta-exotoxin, gamma-exotoxin, and the delta-endotoxin, it is the delta-endotoxin which is of primary interest in registered microbial insecticide products. Title 40 of the Code of Federal Regulations Part 180.1011 requires food use preparations of B. thuringiensis to be free of the beta-exotoxin, called the fly toxin or thermostable exotoxin. The fly bioassay is the method required for detecting the presence of beta-exotoxin. In the 1988 B. thuringiensis Registration Standard, the 40 CFR 180.1011 testing requirements set forth for food use B. thuringiensis products were required for non-food products as well. In addition to the fly bioassay, the Registration Standard required a specific confirmatory method, such as HPLC, for certain storage stability requirements described therein.

B. thuringiensis has been registered as a microbial pest control agent since 1961 under the Federal Insecticide, Fungicide, and Rodenticide Act. It was first registered for susceptible lepidopteran-pests. The product labels listed dosages and representations of insecticidal activity according to the number of spores per gram of product. Since there is no reliable relationship between spore count and insect killing power of a preparation, this resulted in variable potency products and product failures in the field. This measurement of pesticidal activity would have been correct assuming that B. thuringiensis exerted its pesticidal activities by creating an epizootic in the target species. However, B. thuringiensis rarely induces epizootics except in cases where insects are in confined or crowded areas such as insect rearing facilities, stored grain bins, or bee hives (2). The activity of registered B. thuringiensis products is primarily due to the delta-endotoxin which is the major component of the characteristic parasporal crystals. Lepidopteran larvae have been classified into three types based on their susceptibility to crystalline endotoxin, bacterial spores, or mixtures of the two (3). Type I insects are killed by preparations of crystalline delta endotoxin alone and spores of the bacterium do not increase toxicity; Type II insects are susceptible to endotoxin but the effect is enhanced by the presence of spores; Type III insects are killed only by spore-endotoxin mixtures.

The species of B. thuringiensis is divided into serotypes based on antigenic differences of the flagella. To date the species of B. thuringiensis contains 21 serotypes which are subdivided into 36 subspecies. Thirteen of the fourteen distinct crystal protein genes- the so called cry (crystal protein) genes-specify a family of related insecticidal proteins. The insecticidal spectrum of each subspecies differs; however, four major classes have been identified based on the type of delta-endotoxins produced: 1) lepidopteran-specific, 2) dipteran-specific, 3) lepidopteran-and dipteran-specific, and 4) coleopteran-specific. In the three coleopteran-specific B. thuringiensis strains described so far, each of the strains produces crystals producing only one major protein which has been shown by cloning and sequencing to be the same crystal protein in all three strains. In the other three classes the crystals contain more than one major protein. In isolates that have multiple crystal toxin genes the composition of the crystal delta-endotoxins can change in different fermentations of the same isolate. Many strains of B. thuringiensis produce several crystal proteins simultaneously and the same (or very similar) crystal proteins occur in B. thuringiensis strains of different subspecies. This mobility of crystal protein genes among strains of B. thuringiensis subspecies is not unexpected, since most genes are located on large conjugative plasmids.

Under the general labeling requirements of Title 40 of the Code of Federal Regulations Part 156.10, every registered pesticide must have an ingredient statement on its approved label listing the active ingredient(s) and the corresponding percentages of each active ingredient so declared. In 1971 the Environmental Protection Agency issued Pesticide Regulation Notice 71-6 (PR Notice 71-6) which established the International Units (IU) as the basis of expressing potency of products containing B. thuringiensis. PR Notice 71-6 stated that no "quantitative analytical procedures have yet been developed" with regards to the delta-endotoxin. The Agency and U.S. Industry chose to use a standardized bioassay to determine the amount of active ingredient present in each B. thuringiensis product (4). A preparation of B. thuringiensis subsp. thuringiensis, produced by the Institut Pasteur in Paris and designated E-61, was adopted as the international primary standard for bioassay of B. thuringiensis spore-crystal complexes, and was assigned a potency of 1000 International Units/mg. The test insect utilized for the bioassays was the cabbage looper, Trichoplusia ni (Hubner). The use of a standard, assayed concurrently with the samples, serves two purposes. The first is that it helps to correct for variations in insect susceptibility, and assay techniques and conditions that impact on LC_{50}'s, both within a laboratory and between laboratories (5). However, the use of a standard does not always correct for differences in bioassay procedures, such as inclusion of antimicrobials or differences in larval age of the insect species used for testing. The second is that the potencies of B. thuringiensis products for lepidopteran control can be standardized as follows:

$$\frac{LC_{50}\ \text{standard}}{LC_{50}\ \text{sample}} \times \frac{\text{potency of standard}}{1} = \text{potency of sample}$$

Bioassay quality control criteria are an indispensable part of conducting bioassays to reliably determine the killing power of B. thuringiensis preparations, as bioassays are valid only with rigidly standardized assay procedures. Product labels were required to list the active ingredient percentage based on the assumption that 100% product would contain 500,000 IU/mg. Although the active ingredient percentage was determined by bioassay, spore weight percent was required on the label.

Since PR Notice 71-6 was issued, additional B. thuringiensis subspecies have been registered. In addition to lepidopterans, dipterans and coleopterans have been approved as targeted pests on B. thuringiensis labels. Within testing laboratories, new standards have replaced the original E-61 formulation of B. thuringiensis. The E-61 standard is still viable and available from the Institut Pasteur and was used in the standardization of both the 1-S-1971 (6) and 1-S-1980 (5) standards. Very few products or researchers today use isolates of B. thuringiensis subsp. thuringiensis and the newer standards 1-S-1971 and 1-S-1980 were developed using B. thuringiensis subsp. kurstaki. The 1-S-1971, with a potency of 18,000 IU/mg, served as the primary reference standard in the U.S. from 1972 to 1980 for bioassays of lepidopterous-active B. thuringiensis isolates but by 1979 the supply had diminished to a level necessitating the development of a new B. thuringiensis subsp. kurstaki standard. The 1-S-1980 was developed in 1980, with a potency of 16,000 IU/mg, and has served as the kurstaki standard until the present time. The 1-S-1980 was standardized against the E-61 and 1-S-1971, but the publicly held material has lost about 2,000 IU/mg potency and is now bioassaying at about 14,000 IU/mg (Beegle,C.C., U.S.D.A., personal communication, 1990). This material is stored at the USDA Northern Regional Research Center in Peoria, Illinois. There still is some full potency 1-S-1980 in existence in several locations.

There have been four B. thuringiensis subsp. israelensis standards developed for use in bioassay of dipterous-active B. thuringiensis preparations. The IPS-78 and IPS-80 were not stable in storage and were replaced by IPS-82 which has been stable and is available from the Pasteur Institut and has a potency of 15,000 IU/mg. The majority of the 968-S-1983 standard developed has been lost. There are two standardized bioassays for the bioassay of mosquito-active B. thuringiensis preparations. One is the WHO method (7) designed with flexible protocols in view of differing conditions and availability of materials throughout the world. The second is the U.S. standard bioassay (8) with rigid protocols for U.S. conditions. In the U.S. bioassay, early 4th instar Aedes aegypti is specified as the bioassay insect.

Lepidopterous species of Spodoptera are weakly or nonsusceptible to B. thuringiensis subsp. kurstaki or subsp. thuringiensis isolates, and nonsusceptible to B. thuringiensis subsp. israelensis. Until recently there was a lack of an adequate standard. At present there is a B. thuringiensis subsp. entomocidus standard, with a potency of 10,000 IU/mg, for use in bioassays against Spodoptera spp.(9).

Three coleoptera-specific B. thuringiensis strains have been described to date: B. thuringiensis subsp. tenebrionis (10), B. thuringiensis subsp. san diego (11), and B. thuringiensis EG2158 (12). Each of these strains produces rhomboidal crystals containing one major protein. A publicly available subsp. tenebrionis standard has not been developed to date.

The bioassay mandated in PR Notice 71-6 measured the amount of lepidopteran active toxin present via the activity of delta-endotoxin in a bioassay. This measurement of activity was then converted to active ingredient percentage. The December 1988 B. thuringiensis Registration Standard rescinded PR Notice 71-6 and required that the active ingredient declarations be based upon percentage by weight of insecticidal toxin(s) present determined by analytical methods (14). As with products prior to the 1988 Registration Standard, subspecies designation of B. thuringiensis in the product was required. The percentage of delta-endotoxin was required to be declared on the label for each order of insects affected. Potency units were allowed to remain on the label but as an option to registrants. In addition, at this time, the Agency encourages registrants to place specific bioassay information on product labels as an aid to the purchaser.

In conclusion, the historical quantitative requirements on the label for B. thuringiensis were initially for spore counts per gram of product. This was replaced with the requirement to list the active ingredient percentage based on units of potency (IU/mg). Currently the active ingredient percentage is to be based on the percentage by weight of insecticidal toxin(s) present as determined by analytical methods. At present, the Agency is reevaluating the need for potency determinations to be included in addition to the active ingredient percentage.

Literature Cited

1. Himeno, M. J. Toxicol., Toxin Rev. 1987, 6, 45.

2. Dulmage, H.T.; Soper, R.S.; Smith, D.B. In Application of Biorational Substances and Natural Enemies 1st Japan/ USA Symposium on IPM, Tsukuba Kenkyu Gakuen Toshi, Japan, 1981; p. 112.

3. Heimpel,A.M.; Angus, T.A.; J.Insect Pathol. 1959, 1, 152.

4. Dulmage, H.T.; Boening, O.P.; Rehnborg, C.S.; Hansen, G.D.; J. Invertebr. Pathol. 1971, 18, 240.

5. Beegle, C.C.; Couch, T.L.; Alls, R.T.; Versoi, P.L.; Lee, B.L.; Bull. Entomol. Soc. Am. 1986, 32, 44.

6. Dulmage, H.T. Bull. Entomol. Soc. Am. 1973, 19, 200.

7. World Health Organization (WHO) Mosquito Bioassay Method for Bacillus thuringiensis subsp. israelensis (H-14) (Annex 5 in WHO Report TDR, VEC-SWG (5)/ 81.3).

8. McLaughlin, R.E.; Dulmage, H.T.; Alls, R.; Couch, T.L.; Dame, D.A.; Hall, I.M.; Rose, R.I.; Versoi, P.L. Bull. Entomol. Soc. Am. 1984, 30, 26.

9. Salama, H.S.; Foda, M.S.; Sharaby, A. Trop. Pest Manage. 1989, 35, 326.

10. Krieg, A.A.; Huger, A.; Langenbruch, G.; Schnetter, W. J. Appl. Entomol. 1983, 96, 500.

11. Herrnstadt, C.; Soares, G.G.; Wilcox, E.R.; Edwards, D.L. Bio/Technology 1986, 4, 305.

12. Donovan, W.P.; Gonzalez, J.M.; Gilbert, M.P.; Dankosik, C. Mol. Gen. Genet. 1988, 214, 365.

13. Hofte, H.; Whiteley, H.R. Microbiol. Rev. 1989, 53, 242.

14. Registration Standard for the Reregistration of Pesticide Products Containing Bacillus thuringiensis as the Active Ingredient. Case Number 0247, EPA, OPP, 1988.

RECEIVED March 1, 1990

Chapter 3

Bioassay Methods for Quantification of *Bacillus thuringiensis* δ-Endotoxin

Clayton C. Beegle

Insect Pathology Laboratory, Beltsville Agricultural Research Center–West, Agricultural Research Service, U.S. Department of Agriculture, Beltsville, MD 20705

The historical development of *Bacillus thuringiensis* (BT) bioassays in Europe and the United States(US) are traced and the important milestones noted. The 1971 "official" U.S. lepidopterous bioassay and its 1982 revision are covered and bioassay quality control criteria explained. The development of each of the lepidopterous-active standards E-61, 1-S-1971, and 1-S-1980, are discussed. Diet-incorporation *per os* bioassays, surface contamination, Chauthani's hole-in-wax, and Hughes' droplet technique are described for use in lepidopteran assays. Various *in vitro* assays are contrasted with *in vivo* assays and their limitations discussed. The standards and assays of *Bacillus thuringiensis* var. *israelensis* are presented.

History

The first ten years of BT commercialization in the US was marred by product standardization problems. In 1957, Pacific Yeast Products produced a product, Thuricide®, whose fermentation had been optimized based on spore production.This led to the USDA's Pesticide Regulation Division decision to accept spore count as the method to standardize BT products. Since there was (and still is) no reliable relationship between spore count and insect killing power of a preparation (5), this resulted in products with variable potencies and product failures in the field.

Bonnefoi et al. (1) published a description of a standardized procedure for BT preparations in France in 1958. This method used LC50's from a sample and a standard run concurrently in an insect bioassay. This allows the analyst to correct for the daily fluctuations in the standard's LC50. Bonnefoi also mentioned generating "biological units" as a measure of the potency of a preparation based on comparison to a standard.

The first US published BT assay was by Menn (2) in 1960. In 1964, Mechalas and Anderson (3) of Nutrilite Products, first advanced the idea of using a standard and potency ratios in the US. Much of the basic methodology put forth by Mechalas is still used today to standardize BT preparations.

BT Standards

Standards are used to determine relative toxicities for different preparations, correct day-to-day LC50 fluctuations, and to some extent, correct for different assay conditions and methodologies between different laboratories. The measurement of the relative toxicity of a substance by comparing the activity of the substance to a standard has been done since at least 1903 (9).

The first standardization symposium was the 1964 International Symposium on the "Identification and Assay of Viruses and *Bacillus thuringiensis* Berliner Used for Insect Control" held in London, England. Two important resolutions were passed. The first was that spore count was not sufficient for BT standardization; the second recommended the use of LC50's and standard preparations.

The second standardization meeting was the 1966 "Symposia on the Standardization of Insect Pathogens" held in Wageningen, Netherlands. At that meeting it was recommended that a preparation of *Bacillus thuringiensis* var. *thuringiensis* produced by the Pasteur Institute in Paris and designated E-61, be adopted at the international primary standard for the bioassay for BT preparations. E-61 was assigned a potency of 1000 international units (IU)/mg (4).

There have been a number of BT bioassay standards developed since E-61 (Table 1). E-61 is still viable and available from the Pasteur Institute and was used in the standardization of both 1-S-1971 (6) and 1-S-1980 (7). 1-S-1971 was the primary reference standard used from 1972 to 1980 while the 1-S-1980 reference standard is currently used to assess potency of BT lepidopteran-based products. Both 1-S-1971 and 1-S-1980 are powders based on *Bacillus thuringiensis* var. *kurstaki* since very few products or researchers use isolates of var. *thuringiensis* today. The public supply of 1-S-1980 is now held at the USDA Northern Regional Research Center in Peoria,Illinois.Unfortunately, this standard has lost about 2000IU/mg potency and is bioassaying at 14,000 IU/mg (Martinat, P.J., Abbot Laboratories, personal communication, 1989). This potency loss is probably the result of shipping the lot around the US several times between 1986 and 1988.

Salama, et al. (8) have developed a var. *entomocidus* -based preparation, 635-S-1987, for used in bioassays against *Spodoptera* species since the *kurstaki*-based standard is only weakly effective on members of this genus. This standard has an assigned potency of 10,000 IU/mg.

There have been four BT standards prepared for use in assaying dipteran-active products (Table 1). IPS-78 and IPS-80 were not stable in storage and were replaced by IPS-82 which has been stable. IPS-82 is the only dipteran reference standard available from the Pasteur Institute.

Table 1: *Bacillus thuringiensis* Bioassay Standards

Designation	Developer	Variety	Potency(IU/mg)
E-61	Institute Pasteur	*thuringiensis*	1,000
1-S-1971	USDA and industry	*kurstaki*	18,000
1-S-1980	USDA and industry	*kurstaki*	16,000
IPS-78	Institute Pasteur	*israelensis*	1,000
IPS-80	Institute Pasteur	*israelensis*	10,000
IPS-82	Institute Pasteur	*israelensis*	15,000
968-S-1983	USDA and industry	*israelensis*	4,740
635-S-1987	Nat'l.Res.Ctr.,Egypt	*entomocidus*	10,000

Lepidopteran Bioassays

There are essentially four different types of lepidopteran bioassays: diet incorporation, surface contamination, hole-in-wax, and droplet. Forced ingestion methods will not be covered since they are currently not used in determination of potencies.

Diet-Incorporation Bioassays. Diet incorporation is where the BT preparation is mixed into a molten (usually 55 deg. C) agar-based diet. After the diet cools and solidifies, the test insects are placed on the treated diet and the containers covered and incubated for a suitable period of time. Advantages of this universally used method are that there is an even distribution of the

BT sample throughout the diet and that it does not require a high level of either skill or judgement. Disadvantages are it requires a large sample amount, is labor and equipment intensive and cups cannot be set up ahead of time and stored.

There have been two standardized bioassays based on this method. The first was developed by a joint USDA-industry effort (13) and was based on using *Trichoplusia ni* larvae. Although this method served well for two decades, it had two problems: 1) an antibiotic was specified to be used in the semisynthetic diet and 2) too great a faith was placed in the ability of a standard to correct for differences in assay methods. Antibiotics in the bioassay diet can have variable effects depending on the age of the larvae used (14,15). There is no effect when neonate larvae are used but when four-day-old larvae are used, the use of antibiotics results in 3-70 times higher LC50 values depending on the species. It was observed that the shorter the assay time, the greater the effect. Table 2 contains the results from various research labs when they compared identical formulations on diet with antibiotic absent or present. Note that the standard does not correct for larval age.

Table 2: Effect of *Trichoplusia ni* Larval Age and Antibiotic on Potency Determinations of Commercial BT Products at the USDA Brownsville and Upjohn Co.

			USDA Brownsville			Upjohn		
Preparation	Larval Age	Antibiotic Y/N	No. Assays	IU/mg ±SD	%CV	No. Assays	IU/mg ±SD	%CV
SOKW747JF (wettable powder)	Neo	Y	3	13,600 ±814	6	5	11,600 ±1910	16
"	Neo	N	3	15,400 ±1510	10	6	13,400 ±2040	15
"	4day	Y	3	27,700 ±5300	19	--	--	--
"	4day	N	4	30,100 ±5370	18	7	36,300 ±6930	19
SOKL645JB (liquid)	Neo	Y	6	3,640 ±692	19	8	2,950 ±514	17
"	Neo	N	3	3,640 ±369	10	4	3,700 ±534	14
"	4day	Y	3	6,310 ±936	15	--	--	--
"	4day	N	5	7,850 ±1310	17	6	9,420 ±1260	13
Dipel WP 04523BJ	Neo	Y	--	--	--	9	12,300 ±1670	14
"	Neo	N	--	--	--	2	12,100 ±1310	11
"	4day	N	--	--	--	3	24,300 ±1580	6
Dipel WP F-1980	Neo	Y	3	14,800 ±2280	15	--	--	--
"	Neo	N	4	16,300 ±594	4	--	--	--
"	4day	Y	3	12,000 ±2490	20	--	--	--
"	4day	N	4	11,500 ±675	6	--	--	--

Table 3 illustrates the effect of antibiotic, larval age, and insect species on potency determinations of Dipel®. The magnitude of the antibiotic and larval age effects depend on the species assayed.

Table 3: Effect of Larval Age, Antibiotic and Insect Species on Potency Determination of the BT product Dipel

		IU/mg		
Larval Age	Antibiotic in diet?	*Trichoplusia ni*	*Heliothis virescens*	*Ostrinia nubilalis*
Neonate	No	14,100	7,830	11,500
Neonate	Yes	11,300	6,840	13,800
4-day old	No	13,900	11,500	21,800
4-day old	Yes	14,600	15,800	32,300

Modified from Beegle, C.C., Lewis, L.C., Lynch, R.E., and Martinez, A.J., Interaction of larval age and antibiotic on the susceptibility of three insect species to *Bacillus thuringiensis*, J. Invertebr. Pathol., 37, 143, 1981.

Table 4 contains the results of a blind test where differing amounts of viable 1-S-1980 was added to Sandoz and Abbot flowable carriers. These samples along with flowable Dipel and Thuricide samples were coded and sent out for assay to USDA and industry labs. Except for two values (2870 for Thuricide® and 16,100 for Dipel) the results in Table 4 substantiate the observations in Table 2 concerning the commercial formulations. That is, the use of 4-day old larvae in the bioassay results in significantly higher potencies than when neonate larvae are used. It was shown by Beegle et al. (15) that 1-day old *T.ni* larvae respond the same as neonate larvae in BT bioassays. It is noteworthy that larval age did not significantly affect the potencies when 1-S-1980 was incorporated into the commercial flowable carrierrs. This indicates that the observed larval-age difference effect on determined potencies is due to differences in the bacteria rather than formulation. It is also interesting to note the range of IU values determined for neonate and 1 day old larvae for the identical materials when determined by differenct laboratories. There is a relatively narrow potency range for 02-S-1980 (2010-2870 IU/mg) while Dipel exhibits a rather large potency range (3720-16,100 IU/mg).

Table 4: Results of Blind USDA-Industry Bioassay Test

				IU/mg					
Laboratory	Larval Age	No. Assay	CV Range	01-S 1980	02-S 1980	03 Thuricide	04-A 1980	05 Dipel	06A 1980
USDA Brownsville	Neo	18-20	15-18	1700	2610	1290	4170	5630	5620
USDA-Columbia	1day	3-4	--	1660	2010	1380	4210	16100	6620
Sandoz	4day	--	9-19	1960	2760	3970	3310	8110	4520
Abbott	Neo	6-14	5-15	2340	2870	2870	3180	3720	3480
Boyce Thompson	--	--	--	1720	2570	3430	3810	8800	5800

From Lewis, F.B., Report on blind BT sample bioassays, USDA Forest Service, New Hamden, CT, 1982.

A 1982 meeting held between the USDA and producers provided further standardization of the diet incorporation technique to address the inadequacies of the first method. This second procedure specifies that the Brownsville diet will be used without antibiotic or KOH, that sonication in a bath sonicator for 5 minutes @ 3 ml/watt sonicator loading will be used for sample homogenization (16), and that test larvae would be multiply or singly held for 4 days at 30 degrees C.

Surface Contamination Assay. The second most common method, this uses an aqueous sample which has been diluted to the correct concentration and is pipetted onto the surface of solidified diet. The solution is spread evenly over the surface with a glass or metal "hockey stick". When the sample has dried, test larvae are placed on the diet surface, the treatments covered, and incubated. The advantages of this procedure are that it is quick, easy, and cups containing untreated diet can be made up ahead of time and stockpiled. The disadvantages are relatively uneven distribution of the sample, movement of the sample by capillary action down into the juncture of the diet and the side of the container where the diet has pulled away from the cup, and feeding differences between larvae. Larvae which graze along the surface of the diet consume more sample than those who burrow down into the diet in one spot. No commercial companies use this technique to standardize their BT products.

Hole-In-The-Wax Assay. This method was used by Nutrilite Products (10) to standardize their BT product Biotrol®. In this procedure approximately 4 ml of molten diet are dispensed into a 2.5 dram disposable plastic vial. When the media solidifies, a 3 mm O.D. diameter glass rod is inserted about 5 mm into the diet. Molten paraffin is then dispensed onto the surface of the diet to a depth of about 1 cm. After the paraffin has solidified, the rod is removed. The BT sample is introduced into the hold and the vial infested with larvae. The advantages of this technique are that it is conservative of sample, test units can be made up ahead of time and stockpiled, and a uniform dose can be delivered to the test larvae if capillary action between the diet surface and the wax layer does not occur. The disadvantages are that it is expensive, labor intensive, and capillary action can be a problem if diet shrinkage occurs and the surface of the diet pulls away from the wax layer.

Droplet Assay. This ingenious assay was developed by Hughes and coworkers (11). They had observed that newly hatched lepidopterous larvae readily drink water and that the larval uptake of water is very similar between larvae (0.006 ± 0.001 ul) (12). In this technique, neonate larvae are deposited through a funnel onto the benchtop covered with plastic shelf paper. After removal of the funnel, the larvae are left on a small area of the test surface. Then, a rubber stopper on a dowel with pins embedded in the face of the stopper is dipped into an aqueous solution of *B. thuringiensis* and food coloring and pressed onto the surface around the larvae. The larvae surrounded by droplets are then covered by the bottom of a petri dish containing diet. Larvae will imbibe from the droplets, walk into the dish and start feeding on the uncontaminated diet. Larvae that have not imbibed can be identified by the absence of color in their guts and removed from the test. Because the larvae are only briefly exposed to BT, the dosages must be considerably higher than with bioassay systems where the larvae are continually exposed. The advantages of this technique are that it is very conservative of sample and it is quick and easy. The major disadvantage is that when the 1971 and 1980 *kurstaki* standards are compared by this method, the relative potencies are significantly different than when compared by the diet incorporation bioassays. The reason for this disrepency is unknown. Other disadvantages are that some judgement is required when selecting larvae which have imbibed, and there is also some species incompatibility (hairy larvae such as Gypsy moth may become intrapped in the droplets and in darkly pigmented larvae the color change is difficult to detect). Ergo, none of the companies producing BT products use this procedure to standardize their materials.

Dipteran Bioassays

There are two standardized bioassays for assessing the activity of mosquitocidal BT products. The first is the WHO method (17) designed with flexible protocols in view of the differing conditions and availability of materials throughout the world. The second is the US standard bioassay (18) with rigid protocols for US conditions. In the US bioassay, early 4th instar *Aedes aegypti* is specified as the bioassay insect.

Bioassay Quality Control Criteria

Bioassay quality control criteria are an indispensable part of reliably determining the potency of BT formulations. These criteria are of two types, invalidating and discriminatory.

Invalidating Criteria. Presence of any one of these criteria result in invalid assay. They include: 1) greater than 10% check mortality in lepidopteran assays or greater than 5% check mortality in mosquito assays; 2) a nonsignificant F value when an LC50 is calculated on probit analysis; 3) a very atypical slope for the insect species being assayed (usually less than 1 and greater than 4 or 5 depending on the laboratory and the species); 4) over 3 fold difference between the high and low 95% confidence limits of the LC50 (termed very wide confidence limits (VWCL).

Discriminatory Criteria. Assays with shortcomings under these criteria are viewed with suspicion but the results are utilized if appropriate. The discriminatory criteria are: 1) a significant chi-square value is obtained when the LC50 of the assay is determined by probit analysis; 2) there is not at least two mortality points above and below the calculated LC50; 3) there is an atypical slope; 4) erratic dosage-mortality points and 5) 2X to <3X 95% confidence limits (termed wide confidence limits (WCL). For more in depth information on bioassay analysis refer to Finney (22).

Potency Determinations

To determine a mean potency or LC50 value for a preparation it should be bioassayed at least three times over at least two days (three days are preferable). Three values should come from valid assays whose group Coefficient of Variation (CV) is <20%.

When there is an outlier value from a good bioassay, the 4:1 rule is invoked. That is, if there are four values out of five which when taken together the calculated CV is <20%, but when the fifth value is included, the CV is >20%, the outlier value is then discarded.

Potencies are calculated by using the following equation:

$$\frac{\text{LC50 of standard}}{\text{LC50 of preparation}} \times \frac{\text{potency of standard}}{1} = \text{potency of preparation}$$

An area of controversy in the determination of potencies of BT products is in the correction for parallelism. Statistically a potency cannot validly be determined by comparing the LC50 of a preparation to the LC50 of a standard unless there is no significant difference in the slopes of the dosage-response regression analysis, i.e., the slopes must be parallel. Theoretically, the experimental material and the standard can only differ in the amount of toxin, not in the kind of toxin. However, in reality, there are different BT delta-endotoxins and the composition of these toxins can change in relative proportions depending on fermentation and/or formulation conditions. The current solution to this dilemna is to use one of two possible scenarios: 1) for single gene (toxin) products such as the varieties *tenebrionis*, *san diego*, or *kurstaki* isolate 73, use a potency probit analysis program which corrects for parallelism and 2) for multiple toxins and possible formulation effects, ignore the correction for parallelism until an appropriate homologous standard is available.

The aforementioned problem, along with inherent differences in different species sensitivity to various BT products and variation in laboratory potency estimations could contribute to inaccurate predictions of field performance in certain situations.

In Vitro vs *In Vivo* Assays

It has been a dream of BT researchers and producers to replace insect bioassays with chemical or *in vitro* assays. Winkler et al. (19) published a description of an immunochemical technique to determine the amount of crystal protein and its relationship to insecticidal activity as determined by insect bioassay. Their technique took 48 hours to run compared to the 96-168 hours needed to get information from a bioassay. Doing single determinations of 98 samples, they found a 98.3 ± 21% agreement with insect bioassay, and a 107 ± 14% agreement when duplicate determinations were made of 48 samples. Andrews et al. (20) modified the technique and shortened the required time to 4 hours while still maintaining its reliability as an alternative to insect bioassays. However, their conclusions were based on nonreplicated data (toxicity of crystals from one fermentation bioassayed against insects one time). Smith and Ulrich (21) reported on a noncompetitive ELISA technique for quantitative detection of BT crystal protein which takes 4 hours and gives closer agreement to insect bioassay values than does rocket immunoelectrophoresis (RI). They found a mean variation (n=4) of 31% between RI and insect bioassay results, 27% between ELISA and RI, and 6% between ELISA and insect bioassay. Attempts to use SDS-PAGE (sodium dodecylsulfate polyacrylamide gel electrophoresis) and HPLC are also being made to quantitate BT toxins and their relationship to biological activity. A survey of *in vitro* methods as well as specific chapters on HPLC, ELISA, and SDS-PAGE are covered elsewhere in this book.

The current limitations of *in vitro* assays are that they measure the quantity of toxin(s) not the quality. It is the protein quality which determines the insect host range. It is both the quantity and the quality together which determine the killing power of a BT preparation. In this regard, one of the most serious shortcomings of the present chemical methods are their inability to differentiate between undamaged and damaged crystal toxin protein (ie., active vs. inactive protein). Table 5 presents results obtained by Graham Couch (Abbott Laboratories, unpublished data) where they compared toxin quantities and insecticidal activity of three preparations of a *kurstaki* variety by Enzyme Immunoassay(EIA), SDS-PAGE, and insect bioassay. Each sample was split into three subsamples: one subsample was left untouched, one was heated for 5 minutes @ 90 degrees C and the last was heated for 15 minutes at 90 degrees C. The data indicates that while the chemical determinations show no difference between treatments, there is substantial difference in the biological activity.

Table 5: Effect of Heating on *Bacillus thuringiensis* Crystal Toxin Quantity and Biological Actitivy

	Analytical Method		
Treatment	EIA	SDS-PAGE[1]	Insect Bioassay[2]
Nonheated	1	1.0	100%
5 min @ 90degC	1	0.9	7%
15 min @ 90degC	1	0.8	0%

[1]/Relative values

[2]/% Mortality of Trichoplusia ni larvae at 200 ug/ml diet

Chemical methods also cannot measure the presence or absence of spores. There are a number of pest insects that require spores in the BT preparation for maximum toxin activity. Examples are *Colias eurytheme*, *Trichoplusia ni*, *Pseudaletia unipuncta* (23), *Pieris rapae* (24), *Plodia interpunctella* (25), *Laspeyresia pomonella* (26, 27), *Ostrinia nubilalis* (28, 29), *Galleria mellonella* (30), and possibly *Choristoneura fumiferana* (31-33).

Conclusion

This review has attempted to elucidate the integral components of a successful bioassay. The one area left untouched was the insect itself. It is assumed that rearing conditions are uniform and the population is homogenous and well characterized; however, inbred colonies are dynamic and monitoring must be rigorous.

Dr. Rogoff was prophetic over 20 years ago at the 1966 Standardization Symposia, when he stated "Bioassay is valid only with rigidly standardized assay procedures..." Since we are dealing with products produced by living organisms and assaying them on other living organisms the challenge will be to make the information as meaningful as possible by controlling all the other variables as closely as possible. It is clear that at this time there is no substitute for bioassay when assessing the quality of the toxin protein in commercial preparations.

Literature Cited

1. Bonnefoi, A., Burgerjon, A., Grison, P. C.R. Acad. Sci. 1958, 247, 1418-20.
2. Menn, J.J. J. Insect Pathol. 1960, 2, 134-8.
3. Mechalas, B.J., Anderson, N.B. J. Insect Pathol. 1964, 6, 218-24.
4. Burges, H.D., et al.. Mimeo, report of the Second International Symposium on the Standardization of *Bacillus thuringiensis*: Suggested Resolutions, Tests, and Principles, Wageningen, Netherlands, 1966.
5. Dulmage, H.T., Rhodes, R.A. In Microbial Control of Insects and Mites, Burges, H.D., Hussey, N.W. Eds., Academic: New York, 1971, Chapter 24.
6. Dulmage, H.T. Bull. Entomol. Soc. Am. 1973. 19, 200-2.
7. Beegle, C.C., Couch, T.L., Alls, R.T., Versoi, P.L., Lee, B.L. Bull. Entomol. Soc. Am. 1986, 32, 44-5.
8. Salama, H.S., Foda, M.S., Sharaby, A. Trop. Pest Manage. 1989, 35, 326-30.
9. Sun, Y. J. Econ. Entomol. 1950, 43, 45-53.
10. Chauthani, A.R. J. Invertebr. Pathol. 1968, 11, 242-5.
11. Hughes, P.R., van Beek, N.A.M., Wood, H.A. J. Invertebr. Pathol. 1986, 48, 187-92.
12. Hughes, P.R., Wood, H.A. J. Invertebr. Pathol. 1981, 37, 154-9.
13. Dulmage, H.T., Boening, O.P., Rehnborg, C.S., Hansen, G.D. J. Invertebr. Pathol. 1971, 18, 240-5.
14. Ignoffo, C.M., Garcia, C., Couch, T.L. J. Invertebr. Pathol. 1977, 30, 277-8.
15. Beegle, C.C., Lewis, L.C., Lynch, R.E., Martinez, A.J. J. Invertebr. Pathol. 1981, 37, 143-53.
16. Bai, C., Degheele, D. Med. Fac. landbouww. Rijksuniv. Gent. 1984, 49, 1307-14.
17. World Health Organization (WHO) Mosquito Bioassay Method for Bacillus thuringiensis subsp. israelensis (H-14) (Annex 5 in WHO Report TDR.VEC-SWG(5)/81.3).
18. McLaughlin, R.E., Dulmage, H.T., Alls, R., Couch, T.L., Dame, D.A., Hall, I.M., Rose, R.I., Versoi, P.L. Bull. Entomol. Soc. Am. 1984, 30, 26-9.
19. Winkler, V.W., Hansen, G.D., Yoder, J. J. Invertebr. Pathol. 1971, 18, 378-82.
20. Andrews, R.E.,Jr., Iandolo, J.J., Campbell, B.S., Davidson, L.I., Bulla, L.A., Jr. Appl. Environ. Microbiol. 1980, 40, 897-900.
21. Smith, R.A., Ulrich, J.T. Appl Environ. Microbiol. 1983, 45, 586-600.
22. Finney, D.J. Probit Analysis, 3rd edition. University Press: Cambridge. 1971.
23. Somerville, H.J., Tanada, Y., Omi, E.M. J. Invertebr. Pathol. 1970, 16, 241-8.
24. Soliman, A.A., Afify, A.M., Abdel-Rahman, H.A., Attwa, W.A. Anz. Schaedlingsk. Pflanzenschutz. 1970, 43, 116-25.
25. McGaughey, W.H. J. Econ. Entomol. 1978, 71, 687-8.
26. Roehrich, R. Coll. Int. Pathol. Insectes, Paris. 1962,312-3.
27. Vervelle, C. Ann. Parasitol. Hum. Comp. 1975, 50, 655.
28. Mohd-Salleh, M.B. Ph.D. Thesis, Iowa State University, Ames, 1980.
29. Sutter, G.R., Raun, E.S. J. Invertebr. Pathol. 1966, 8, 457-60.
30. Burges, H.D., Thomson, E.M., Latchford, R.A. J. Invertebr. Pathol. 1976, 27, 87-94.
31. Yamvrias, C., Angus, T.A. J. Invertebr. Pathol. 1970, 15, 92-9.
32. Smirnoff, W.A., Valero, J.R. Can. Ent. 1979, 111, 305-8.
33. Fast, P.G. Can. Ent. 1977, 109, 151-508.
34. Wilcox, D.A., Shivakumar, A.G., Melin, B.E., Miller, M.F., Benson, T.A., Schopp, C.W., Casuto, D., Gundling, G.J., Bolling, T.J., Spear, B.B., Fox, J.L. In Protein Engineering Applications in Science, Medicine and Industry, Inouye, M., Sauna, R., Eds., Academic: New York, 1986, Chapter 25, p 407.
35. Kronstad, J.W., Schnepf, H.E., Whiteley, H.R. J. Bacteriol. 1983, 154, 419-28.

RECEIVED March 1, 1990

Chapter 4

Specificity of Insecticidal Crystal Proteins

Implications for Industrial Standardization

R. Milne[1], A. Z. Ge[2], D. Rivers[3], and D. H. Dean[3,4,5]

[1]Forest Pest Management Institute, Sault Sainte Marie, Ontario, P6A 5M7 Canada
[2]Department of Microbiology, Ohio State University, Columbus, OH 43210
[3]Department of Entomology, Ohio State University, Columbus, OH 43210
[4]Department of Biochemistry, Ohio State University, Columbus, OH 43210
[5]Department of Molecular Genetics, Ohio State University, Columbus, OH 43210

The specificity of *Bacillus thuringiensis* activity resides largely with the insecticidal crystal proteins. The "specificity domain" of the toxin moiety has recently been identified and is proposed as the receptor binding domain. The bacterial spore and other factors may play a role in insecticide specificity, especially for less susceptible insects. Insect susceptibility is dependent on gut pH, proteases, the presence and type of receptor (toxin-binding) protein, and on membrane interactions with the cytolytic domains of the toxins. The implications for industrial standardization of *B. thuringiensis* products is that assay systems must account for each of the bacterial components which play a role in insecticidal activity.

Historical Overview of BT specificity. The mode of action of *Bacillus thuringiensis* on insects is dependent upon two factors: Specificity of the microorganism and the susceptibility of the host insect. Recent developments in this area have come from studies on molecular genetics and biochemistry of the *B. thuringiensis* insecticidal crystal protein (ICP) toxins and the interactions of these with insect tissues at the membrane level. Results from these studies allow the proposal of a testable model for the mode of action of ICP toxins. We shall refer to this model as the Wolfersberger/Ellar or Receptor/Pore Model. Although insufficient supporting data has been gathered at the present time to completely verify the model, particularly on the subject of insect susceptibility, we may now speak with more confidence about the first of the two factors of *B. thuringiensis* mode of action, specificity of ICP toxins.

The Model: Wolfersberger, et al. (1) and Ellar, et al. (2) have independently proposed: Proteolytically activated toxin binds to specific

0097–6156/90/0432–0022$06.00/0

receptors on midgut epithelial cells and then forms a pore which disrupts potassium ion flow, eventually causing cell lysis.

The last major review on the subject of specificity and susceptibility was almost 20 years ago (3), but it remains accurate in its perceptions of the key factors in mode of action. The purposes of the present review is to summarize past and recent developments, to address and criticize the Model and to discuss the ramifications of ICP specificity on quantitation of active components of commercial *B. thuringiensis* preparations.

The activity of microbial pesticides is now viewed as dependent upon the variety of toxins produced by the microbe as well as the properties of the host insect. In the early years of research, however, it was believed by some that *B. thuringiensis* was not very specific:

> "The nonspecificity of *B. thuringiensis* is, without doubt, the result of its ability to produce several materials toxic for insects. Although this fact enhances the value of this microbial agent, it also creates problems associated with production and standardization" (4)

Interestingly, we are still addressing these same issues in the present volume and although we have gathered more data on some aspects of specificity and susceptibility, others remain enigmatic.

The variety of toxins present in *B. thuringiensis* were originally identified by Heimpel (4) as alpha-exotoxin (lecithinase C or phospholipase C); alpha-exotoxin (thermostable toxin or fly factor); and delta-endotoxin (the crystal protein). The latter was originally defined as the crystal protein and thought to represent a unique toxin (4). Variability in insecticidal activity was considered to be due to the susceptibility of the insect and three general types of insects were identified:

> Type I - insects which exhibit general paralysis and which show blood pH change to crystal protein alone; e.g., *B. mori*, *M. sexta*
>
> Type II - insects which are susceptible to crystal alone, suffer gut paralysis but no gut leakage, no change in blood pH, or general paralysis; e.g., most *Lepidoptera*
>
> Type III - insects which require both crystals and spores for pathogenesis; e.g., *Anagasta kuehniella* or *Lymantria dispar*. (5)

The classification of insects according to their crystal susceptibility should be reconsidered as the spectrum of activity of *B. thuringiensis* insecticidal crystal protein become more available. Indeed, sufficient data existed by 1956 to show that different strains of *B. thuringiensis* exhibited a host species susceptibility spectrum, *B. mori*, for example, was shown to be highly susceptible to certain *B. thuringiensis* serotypes but not others (6). In the extremely prophetic early review of the subject of *B. thuringiensis* host spectrum, Burgerjon and Martouret (1971) summarized six general characteristics of host spectrum:

> 1. For each insect species, the degree of entomopathogenic activity expresses the susceptibility to a particular bacterial strain; i.e., an insect species exhibits a spectrum of sensitivity to different *B. thuringiensis* strains.
>
> 2. Different strains of *B. thuringiensis* exhibit different degrees of entomopathogenic activity on the same host species; i.e., a *B. thuringiensis* strain exhibits a spectrum of insecticidal activity.

3. Most species of susceptible insects belong to the orders *Lepidoptera*, with a few species in the orders *Hymenoptera*, *Coleoptera*, *Orthoptera*. [We may now add *Diptera* to this list.]

4. Bacterial strains belonging to the same flagellar serotype have their own specificity spectrum, independent of the serotype.

5. The ED_{50} of different bacterial strains can appear in different sequences for different insect species, so that there are reversals of the sequences for the pair of species, termed "cross activity".

6. The activity spectra of bacterial strains are not correlated with the systematic relationships of the insect species.

It was clear to Burgerjon and Martouret (3) that the bacterial factors of *B. thuringiensis* specificity were the spore, the crystal, and the thermostable exotoxin (beta-exotoxin or thuringiensin). Other factors, such as proteases, phospholipase, and other proported toxins mentioned above are not present in crystal-spore preparations, but may play some role in bacterial specificity in those cases where bacterial pathogenesis and septicemia take place. These other factors were not strongly documented in 1971 and have not received significant attention recently. We shall not consider beta-exotoxin in this review since it is not present in most commercial preparations and therefore not a factor in specificity.

The factor that was not readily apparent in 1971 was the diversity of specificity due to the ICP's alone. While it was recognized that "the crystals from various strains of *B. thuringiensis* exhibit appreciable differences in toxicity against the same host species..." (3), it was not until several important studies came to light that the diversity of insecticidal activities could be associated with the set of ICPs that appeared to be virtually identical. Again, Burgerjon and Martouret (3) noted "the amino acid composition of the crystals of *B. thuringiensis* varies so little between the different strains that it does not seem possible that such slight variations could be responsible for the pronounced toxicity differences." They proposed, "the differences in toxicity can be explained by the manner in which the *in vivo* hydrolysis is effected...[suggesting] differences in the structure of the crystals, some of whose bonds might be less accessible to the insect enzymes." It has taken 20 years to provide supporting data for this prophetic conclusion, but an even more exciting prospect went unrealized, the role of toxin-receptor binding in determining specificity.

Much of the progress which resulted in the areas of molecular genetics and cell biology of *B. thuringiensis* mode of action can be traced to the international cooperative effort led by Howard Dulmage, Denis Burges and Hugette deBarjac and sponsored by the World Health Organization (7). This effort led to considerable information on strain specificity [most of which has yet to be published] and perhaps more importantly to a set of strains, the HD collection, which has become a tremendous resource for genetic, biochemical and bioassay data.

Immunological work by Krywienczyk (8-11) revealed that *B. thuringiensis* exhibited different crystal types, which were often mixed and which were not necessarily consistent with flagellar serotypes. This forecasted the genetic diversity of *B. thuringiensis*, but the definitive proof that it could encode several separate ICP genes, singly or in combination, came with the publication of DNA hybridization studies by Kronstad, Schnepf and Whiteley (12).

The above historical overview has laid the foundation for the following review which shall address in greater detail the questions first raised by Burgerjon and Martouret (3). What are the bacterial components which cause specificity differences seen among *B. thuringiensis* strains, and what are the host insect components which affect the susceptibility to *B. thuringiensis*.

Role of Bacterial Components

Role of the insecticidal crystal proteins.

As a first approximation, we may say that the insecticidal spectrum of each *B. thuringiensis* strain is determined by the type of crystal protein genes it carries, and possibly, the extent of how well these genes are expressed *in vivo*. But even this tenet will have to be modified in some insects when we consider the role of the spore and synergy of different toxins. In a recent review, a standardized nomenclature was proposed (13). According to this nomenclature, the currently known crystal proteins can be classified into four major groups based on their specificity and homology as well as protein structure. CryI proteins are specific against *Lepidoptera*; CryII proteins are specific to both *Lepidoptera* and *Diptera*; CryIII proteins are effective against *Coleoptera*; and CryIV proteins are specific to *Diptera*. Proteins within the CryI group have extensive sequence homology ranging from 55% to 90%. Sequence homology among different groups seems to be equivalent to or less than 60%. Nevertheless, it is possible to identify five highly conserved amino acid blocks among most of the crystal proteins (13).

Of CryI type protein toxins, CryIA(a), CryIA(b), and CryIA(c) are undoubtedly the best studied proteins. At present, more than 20 *B. thuringiensis* strains of ten serotypes are found to carry at least one of these three genes (13). Even though these three proteins, CryIA(a,b,c), share more than 80% sequence homology, the specificities of these proteins are markedly different in the cases of *B. mori* and *H. virescens* (Table I), in the case of *B. mori* and *H. virescens*.

An alignment of amino acid sequences of CryIA(a,b,c) reveals that the major difference between CryIA(a) and CryIA(b) lies between amino acids 340 and 450, and the difference between CryIA(b) and CryIA(c) is within amino acids 467 to 612. CryIA(a) differs from CryIA(c) over the region from 340 to 612. This region has been referred to as the hypervariable region (Geiser, Schweitzer and Grimm (16) and has been shown to be the region of the protein associated with the specificity of the toxin.

Location of Specificity Domains. Because of the lack of knowledge about the structure of *B. thuringiensis* insecticidal crystal proteins, past discussions about the specificities of crystal proteins tend to emphasize host processes which act upon crystal proteins rather than intrinsic properties of proteins themselves. Aside from host factors such as insect midgut proteases, pH, and types of crystal protein toxin receptors, the specificities of insecticidal crystal proteins seem to be determined by amino acid sequences located on the variable domains of these proteins. In an experiment by Ge, et al. (15), DNA sections of the hypervariable region of *cryIA(a)* and *cryIA(c)* were exchanged to test the change of protein specificity. The block exchange experiment has

Table I. Specificities of CryIA Type Protein Toxins

	LC_{50} of solubilized crystal proteins of			
Lepidoptera	CryIA(a)	CryIA(b)	CryIA(c)	Reference
P. brassicae[a]	0.8 (0.5-1.4)	0.7 (0.6-1.1)	0.30 (0.1-1.3)	13
M. brassicae[b]	0.17 (0.12-0.23)	0.162 (0.09-0.28)	2.00 (nd)[d]	13
C. fumiferana[e]	0.052 (0.031-0.088)	nd	1.80 (nd)	14
S. littoralis[b]	>1.35 (nd)	>1.35 (nd)	>1.35 (nd)	13
B. mori[c]	0.37 (0.001-0.73)	nd	>150	15
B. mori[e]	0.028 (0.01-0.04)	nd	2.67 (2.17-3.40)	14
T. ni[c]	2.9 (1.34-5.67)	nd	0.32 (0.12-0.81)	15
M. sexta[b]	0.0077 (0.024-0.135)	0.0086 (0.006-0.013)	0.0072 (0.06-0.08)	13,15
H. virescens[b]	0.090 (ng[f])	0.010 (ng)	0.0016 (ng)	13

[a]Concentration at ug/ml
[b]Concentration at ug/cm^2
[c]Concentration at ug applied to artificial diet or leaf disc.
[d]nd: not determined
[e]Force-fed concentration in ug/ml
[f]ng: not given

demonstrated that the *Bombyx mori* specificity region is located within amino acids 332 to 450 on CryIA(a) (15). The exchange does not seem to distort the overall protein functional structure as the new hybrid protein possesses toxicity against *B. mori*. It is possible that 118 amino acids defined as *B. mori* specificity domain on CryIA(a) constitute a specificity domain on the protein toxin. After acquiring this specificity domain from CryIA(a), CryIA(c) is highly toxic to *B. mori*. We have speculated that the specificity domain of the *B. thuringiensis* ICP identified in these experiments is the receptor binding domain (15).

If the amino acid sequence 332 to 450 on CryIA(a) does constitute a binding domain to receptors on the brush border membrane of *B. mori* midgut epithelial cells, then the concept of a binding domain may apply to variable regions of other crystal proteins. Hofmann, et al. (17) have shown that activated CryIA(a), CryIA(b), and CryIA(c) toxins bind with high affinity to receptors on the brush border membrane of *Manduca sexta* midgut epithelial cells, and the binding is competitive indicating that all three toxins share the same type of receptor (17). The three toxins, however, have different 50% binding saturation concentrations reflecting their different LC_{50}. It is possible that, even though the binding domain to receptors of *M. sexta* is present in each toxin, the affinity between the binding domain and the receptor varies among three toxins depending upon the amino acid sequence. A low LC_{50} value may signal a better binding group on a crystal protein.

The high degree of variation in sequence of amino acids among crystal proteins can be useful in specificity differentiation. Hofte, et al. have analyzed the insecticidal spectra of CryIA, CryIB, and CryIC, using monoclonal antibody (MAB) (18). Their data reveals that only those MABs that react with epitopes formed within the variable domains are highly unique and capable to further differentiate specificities of crystal proteins. However, caution should be made when one tries to compare MABs prepared from denatured and native crystal proteins. As a powerful tool of differentiation of specificities of crystal proteins, MABs should be prepared from native crystal proteins, eliminating factors that are likely to alter natural epitopes.

Role of the Spore. The bioassay of delta-endotoxin using susceptible lepidopteran larvae, for the purpose of defining specificity, requires techniques which yield both known dosage and subsequent response. The known dose is easily achieved by force-feeding. However, the response can vary from a reduction in feeding or larval weight gain to mortality. If the preferred response is mortality, then difficulties may be encountered with certain larval types as pointed out by Heimpel and Angus (5). They indicated that a Type III larva required the presence of spores for a mortality response, whereas Type I and Type II do not require spores.

The importance of spores in enhancing the mortality response in *Galleria mellonella* has been investigated by Burges, et al. (19). They showed that increased mortality was related more to the spore concentration than to crystal. Fast (20), however, reported that spores alone played little or no role in the mortality of spruce budworm, *Choristoneura fumiferana* (Cf). Our experience performing similar bioassays on Cf using purified crystals, filter-sterilized protoxin or toxin, and more recently, inclusion bodies from recombinant *E. coli*, often resulted in highly variable mortality. We have observed that these larva will exhibit a feeding inhibition response to spore-free toxins, but can easily recover and resume normal weight gain through to pupation. It appears that damage to the gut causing feeding inhibition can be repaired within a few days. Regeneration of midgut cells in response to toxin damage has previously been reported by Spies and Spence (21) for *Manduca sexta*.

We reasoned that the gnotobiotic character of Cf larvae reared on diet containing antibiotics may, in part, be responsible for the variable mortality response. In order to test this hypothesis, purified spores from the HD-1-S-80 standard were included in serial dilutions of *E. coli* inclusion bodies and a 2 ul dose was force-fed to early sixth instar Cf larvae. Renografin purified spores (22) were further washed for 24 h in 0.05 M KOH with two changes of alkali to remove contaminating crystal protein. A viable spore count was determined after washing. Inclusion bodies from the two gene types, designated 4101 and 4201, were assayed in parallel using water or spores as the diluent, and a spore-only dose was included as control. The level of spores in a 2 ul dose was calculated to be 2.7 X 103. Mortality was scored for each of seven days, and LD_{50} for treatments was computed by probit analysis. The data, presented in Table II, shows a reduction in LD_{50} for comparable days when spores were included. More significantly, the time to reach an LD_{50} response was reduced by 5-6 days. It appears that relatively high dose levels of toxin are not lethal in these "gnotobiotic" larva unless an opportunistic infection can be established.

The variability in response to sterile toxins from both native and recombinant *E. coli* can be partially controlled by ensuring each larva ingests a dose of spores sufficient to initiate septicemia. Although relying on a secondary effect for mortality, the technique of spore enhancement overcomes the often heterogenous mortality response seen in larval bioassays.

Pertinent to the discussion of the role of the spore on specificity is the the comparison of CryIA(a) with CryIA(c) activity on spruce budworm, with and with out spores, shown in Table II. Without spores, the CryIA(a) crystal type is 23 X more toxic than CryIA(c) (day 7). However, with spores, the opposite observation is seen. CryIA(c) is 3X more toxic. The role of the spore changes the specificity ratio of the two ICPs. These observations certainly deserve further examination, but clearly the spore is not merely an inactive spectator in the specificity of ICP action on Type III insects. Similar experiments were also conducted on *B. mori* (a Type I insect). For this insect the specificity ratio between CryIA(a) and CryIA(c) toxins remained the same, with or without spores.

We feel the "with spores" case is a more accurate reflection of the toxic action in that the response is observed within 24-48 h. This provides a closer link in time between dose and response. The rapid mortality seen in these laboratory assays is in keeping with the observation that serial application of commercial formulations of *B. thuringiensis* for spruce budworm control routinely results in 24 h mortality. It is worthy to note that the addition of spores yields a qualitative effect in addition to a quantitative one.

Role of the Insect

Solubilization of the Crystal/Gut pH. Crystal proteins of *B. thuringiensis* are normally insoluble in neutral solutions [except under conditions where cytotrophic salts are used (Fast and Milne, 1979, J. Invert. Pathol. 34:319-) and require an alkaline environment and proteases for conversion from the crystal to the protoxin and to the activated toxin *in vitro* (23). Insects susceptible to *B. thuringiensis* crystal proteins would presumably possess an alkaline gut environment with the necessary proteases to activate the toxin. An absence of either factor in the insect gut should reduce the susceptibility of that insect toward *B. thuringiensis* crystal proteins. We shall discuss these possibilities in this section by first examining the conversion of crystal proteins to activated toxin in the guts of susceptible insects and then focusing on insects that display lower susceptibility to *B. thuringiensis* crystal proteins.

An alkaline gut pH is characteristic of lepidopteran insects that feed primarily upon plant material (24). The alkaline pH aids in the breakdown of tannins and other plant material that would be insoluble at lower pH ranges. This alkaline pH is also responsible for the conversion, upon ingestion by the insect, of *B. thuringiensis* crystal proteins into the lower molecular weight protoxin (25,26). The protoxin is further transformed to the activated toxin by action of proteases within the gut of the insect (27). However, this conversion to the activated toxin cannot occur without the alkaline transformation of crystal protein to the protoxin.

Lecadet and Dedonder (28) demonstrated that two serine proteases isolated from the midgut of *Pieris brassicae* could convert *B. thuringiensis* Berliner crystal protein to the activated toxin in the absence of alkaline-reducing agents. Later, Faust et al. (26) demonstrated that the proteases and the alkaline storage buffer used for these enzymes were both required for the conversion of crystal protein to activated toxin, and further

Table II. Comparison of the LD_{50} response of *C. fumiferana* to per os microinjection of crystals and crystals containing *B. thuringiensis* spores

	LD_{50} ng/larva								
	DAYS[1]								
Toxin	1	2	3	4	5	6	7	8	9
CryIAa + water	a	a	a	a	a	78 (42-145)	52 (31-88)	b	b
CryIAa + spores	82 (43-260)	28 (-)	12 (6-20)	8 (1-11)	c	c	c	c	c
CryIAc + water	a	a	a	a	750 (-)	1800 (-)	1250 (-)	b	b
CryIAc + spores	28 (18-48)	12 (4-21)	8 (-)	6 (1-10)	c	c	c	c	c
spores only	--------------no mortality----------------								

[1] no readings taken on Day 2

a Insufficient mortality to calculate LD_{50}

b Data too heterogenous for analysis

c Termination of assay; larva beginning to pupate

(-) indicates that confidence limits were not available.

showed that an alkaline environment was required prior to protease activation. This has been supported by Tojo and Aizawa (23) who demonstrated that *Bombyx mori* gut juice could not solubilize crystal proteins from *B. thuringiensis kurstaki* (HD-1) if the pH was adjusted below 10.0. They further demonstrated that alkaline solubilization does proceed protease activation. Further investigations have demonstrated that susceptible lepidopterans must have an intestinal pH greater than 8.9 for successful conversion of crystal to protoxin to activated toxin (5,6,29). There are exceptions to this, as noted for *Mamestra brassicae* (pH 10.2), which was found to be unable to solubilize the crystal protein of *B. thuringiensis* Berliner (29). If confirmed, this would suggest that the structural composition of crystal proteins dictates the degree of solubilization in a given insect gut environment.

From the discussion above, we can conclude that in the gut of susceptible lepidopteran insects, crystal proteins are converted to a protoxin by the alkaline pH of the gut juices, and further transformed into the active toxin via the action of gut proteases. This appears to be the case for susceptible insects, but what is happening in the gut of insects that display lower susceptibility? Yamvrias (30) demonstrated that the Mediterranean flour moth, *Anagasta kuehniella* (neutral gut pH), was not able to solubilize the crystal protein of *B. thuringiensis* Berliner and was therefore not susceptible to the toxin. The inability to transform the crystal to protoxin is believed to be due to the low gut pH of *A. kuehniella*, which is near neutrality. Similar observations have been noted for the European corn borer, *Ostrinia nubilalis*, and the corn earworm, *Heliothis virescens*. Rivers and Dean (unpublished) and Rivers et al. (31) demonstrated that *O. nubilalis* was not able to solubilize the crystal protein of *B. thuringiensis kurstaki* (HD-73) owing to its near neutral gut pH (6.4-7.8), and thus displaying low susceptibility to the crystal toxin. Jaquet et al. (25) demonstrated poor transformation of several *B. thuringiensis* crystal proteins to protoxin in the gut of *H. virescens* (pH 8.5). Again, this insect possesses a gut pH below 8.9 that was alluded to earlier as characteristic of *B. thuringiensis* susceptible insects. Although many lepidopteran insects display low susceptibility to *B. thuringiensis* crystal proteins, apparently due to a gut pH not favorable for *in vivo* solubilization, *in vitro* transformation prior to ingestion by the insect may increase activity of crystal proteins toward these insects.

In the case of *A. kuehniella*, susceptibility was demonstrated toward *B. thuringiensis* Berliner crystal proteins when applied with spores, or if dissolved and activated prior to ingestion (30). *H. virescens* is very susceptible to *B. thuringiensis* toxins (LC_{50} = 90 ng, 13) if the crystal is solubilized prior to ingestion (25). Similar observations have been noted for the European corn borer (31), however, solubilization does not increase activity of some crystal proteins. Crystal protein of *B. thuringiensis aizawai* (HD-249) did not display activity toward *O. nubilalis* despite dissolution and activation prior to ingestion (Rivers & Dean, unpublished). *P. brassicae* has been shown to display the same activity to crystal, protoxin, and toxin from several *B. thuringiensis* strains (25). This suggests that although conversion of crystal proteins to the activated toxin via a protoxin is essential for susceptibility of certain insects, other factors such as binding efficiency, proper gut receptors, composition of the protein, etc. are involved in regulating the specificity of these crystal proteins.

Differential proteolytic processing/protease-inhibition of toxin. Lecadet and Martouret (32) have shown that enzymatic hydrolysis is required to activate protoxin. The term "activation" is loosely defined as the enzymatic process which converts the ~130 kDa protoxin to the biologically active toxin normally in the range of ~50-70 kDa. This activation process can be accomplished *in vitro* by endogenous proteases (33), digestive enzymes from the insect gut (34), or with a variety of bacterial, plant, or mammalian enzymes (35).

Prolonged incubation of crystals in alkaline buffer alone results in a stable toxin being produced (33,36,37). Specific enzyme inhibitors have been used to slow this activation, implicating endogenous enzymes in the process (38). The endogenous enzymes have been identified as serine-, leucine amino-, and metallo-proteases (33,39). Andrews et al. (36) have also shown that a *B. thuringiensis* strain with reduced intracellular proteases does not autolyse the crystal, offering further evidence that these enzymes can be responsible for activation of toxin. The products from these digestions have not been characterized as to where the enzymes cleave the protoxin, and little is known of what role these enzymes play in the *in vivo* activation of toxin. The *Lepidoptera* which show susceptibility to delta-endotoxin have, in general, high alkaline midgut digestive systems with characteristically high pH optima for their trypsin-like and chymotrypsin-like digestive enzymes (40,41,42). The activation of crystals in the undiluted gut juice from *P. brassicae* occurs in 20 seconds (43). This rapid activation has also been observed for several crystal types incubated in the neat gut fluids from *B. mori*, *T. ni*, and *C. fumiferana* (H. Falter and L. Burns, Laurentian University, unpublished). Some investigators have partially purified the gut enzymes from different larvae (44,45,46) and shown subsequent activity in the processing of protoxin. However, the resultant toxin fragments have only been characterized by molecular weight and activity.

The activation process for crystals incubated with bovine trypsin has been described by Chestukhina, et al. (33) (Choma, et al., 1990, Eur. J. Biochem., in press) as a step-wise degradation of the C-terminal portion of protoxin resulting in a protease stable fragment. If the incubation is performed at dilute concentration of enzymes, this process appears to hold true. There is also an N-terminal cleavage of approximately 28 amino acids as reported by Nagamatsu, et al. (47), Aronson, et al., (48), and Bietlot, et al. (49) which may be the result of higher enzyme activity during activation. In only one report (49) has the toxin generated by trypsin from HD-73 been characterized by both the N- and C-terminal sequences, offering confirmation of the gene-deduced sequence and delineating an activated toxin fragment.

In the case of the dual specificity of lepidopteran/dipteran toxins, the exact position of the N- and C-termini has been shown to be critical in determining toxicity and dual specificity of toxin. Haider and Ellar (50) showed that for the strain *aizawai* IC1, the toxin generated by lepidopteran enzymes retains lepidopteran activity. This same toxin can be further processed by dipteran enzymes to produce a dipteran specific toxin. In this case, the removal of ~15 amino acids from the C-terminal end of the toxin fragment resulted in a shift in the host specificity. The importance of a characterized N-terminal site has been shown by Hofte et al. (51) using gene deletion and expression. They determined that toxicity was lost in a deletion between amino acid positions 29 and 37.

Hofte and Whiteley (13) have reported toxicity data for six representative single gene products against five larval types. They observed that crystal type influenced the spectrum of activity more so than the species of larva. Jaquet, et al. (25) have previously reported that predissolution, or *in vitro* activation, of protoxin with trypsin alters the activity and specificity of certain crystal types. In neither study has the toxin fragments been characterized as to their cleaved sites, making it difficult to assess whether differential proteolysis is responsible for the quantity of toxin produced, or if the specificity of toxin had been altered. We have observed that toxin fragments show different apparent molecular weights in SDS PAGE analysis, depending on the enzyme system used to generate the fragment. Furthermore, what appears to be a single band on SDS polyacrylamide gels often yields three or more bands when subjected to isoelectric focusing, indicating further processing of either the N- or C-terminal positions. Of concern is whether these minor processing differences would be reflected in altered activity or specificity.

Further studies concerning the differential proteolytic processing of delta-endotoxin require careful attention to quantification and characterization of the activated toxin before conclusions can be drawn concerning the role that larval digestive enzymes play in conferring specificity.

Receptor binding. A potential determinant for insecticidal specificity is the toxin-binding receptor. Receptors, at least for the CryIA proteins, are themselves proteins which are located on the brush border membrane or microvilli of the midgut epithelium. A binding assay using ^{125}I labeled toxin was first employed by Hofmann, et al. (52). Insecticidal specificity of two toxins (CryIAb and CryIB) on two insects (*M. sexta* and *P. brassicae*) showed a correlation between insecticidal activity (specificity) and high affinity binding sites on the respective brush border membrane vesicles (BBMV) (17).

A clearer demonstration of the direct relationship between toxin specificity and receptor binding has recently appeared (53). Comparing LC_{50} (bioassays) with binding to BBMV from *M. sexta* and *H. virescens* using ^{125}I labeled toxins of all three CryIA toxins, insecticidal specificity was correlated with binding affinity (Kd) but even more so to the concentration of binding sites (Ri) (Table III). CryIA(b) and (c) proteins have similar insecticidal activity and receptor binding parameters against *H. virescens*. CryIA(a) is two orders of magnitude less active against *H. virescens* and has both a lower receptor concentration and a weaker binding affinity against BBMV from this insect.

TABLE III. Insect Specificity and Midgut Receptor Binding of CryIA Toxins

Toxin	*M. sexta*			*H. virescens*		
	LC_{50}	Kd	Ri	LC_{50}	Kd	Ri
CryIAa	20	1.1	9.9	157	0.8	3.7
CryIAb	20	0.2	7.0	7	0.4	21.0
CryIAc	9	0.2	6.3	2	0.4	62.3

LD_{50} values ng toxin applied/cm^2 of artificial media.
Kd dissociation constant nM.
Ri pmol/mg BBMV protein. Kd and Ri calculated from homologous competition experiments. Data from Van Rie, et al. (53).

As a group, these papers have built a strong case in favor of toxin-binding to midgut receptors as the major factor in specificity of insecticidal activity of *B. thuringiensis* delta-endotoxins against lepidopterans. Even taking the data in no more general terms, specificity is apparently determined by receptor binding in *M. sexta*, *P. brassicae*, and *H. virescens*.

An apparent exception to this is the recent results of Wolfersberger (54) on *L. dispar* which indicates an inverse relationship between binding affinity (Kd) and specificity. CryIAb protein is 400 times more active than the CryIAc protein (LC_{50} values of 1.08 and 425.0 respectively) while binding affinity is ten times less for CryIAb than for CryIAc (Kd is 19.8 nM versus 2.03 nM respectively).

Implications for Industrial Standardization

The recognition that insecticidal crystal proteins are primary agents in insecticidal activity of *B. thuringiensis* toward certain insects opens the way for the development of simplified assay conditions. These assays could be based on ELISAs using monoclonal antibodies to individual insecticidal crystal proteins, or other assays of expressed insecticidal proteins. These assays could reduce, but not replace, costly and laborious insect bioassays. Thus, commercial formulations may be labeled with units or ug amounts of active ingredient, just as with chemical agents.

An important caveat to the concept of protein assays of insecticidal crystal proteins is that insecticidal crystal proteins are highly specific. A particular insecticidal crystal protein may be highly active against certain insects, while having virtually no activity against other insects. Certain crop pests may show little activity to any known insecticidal crystal protein. Furthermore, combinations of insecticidal crystal proteins, occurring naturally in *B. thuringiensis* isolates or reformulated, may not yield biological activity expected from the sum of the activities of the component insecticidal crystal proteins. It is clear from this warning that protein assay or ELISA assay determination of insecticidal crystal protein content of a product has no inherent relationship to the biological activity of the product against any particular insect unless that that relationship has been previously established. Such product/insect relationships also do not predict the relationship of the product to another insect, nor does one such relationship predict the relationship if the formulation of the product changes (including additions of other insecticidal crystal proteins or chemical agents which change the way that the product might interact with the same or other insects).

With these conditions in mind we may note that the broad range of insecticidal activity of reported for *B. thuringiensis* is now better understood as the result of combined activity of individual insecticidal proteins, each of which has a specific range of insecticidal activity. This understanding, indeed, provides the advantage of employing analytical methods, as described in the remaindure of this book, for detecting and quantitating these proteins.

Acknowledgments

We wish to thank Terry Wright and Christine Budau for excellent technical assistance. The research reported in this review was supported by NIH grant RO1 AI29092 to D.H.D. and by the Forestry Canada, Sault Ste. Marie, Canada to R.M., through the Canadian Forestry Service, Sault Ste. Marie, Ont.

Literature Cited

1. Wolfersberger, M.; Hofmann, C.; Luthy, P. Zentralbl. Bakteriol. Mikrobiol. Hyg. I Suppl. 15, 1986; p. 237.
2. Ellar, D.J.; Thomas, W.E.; Knowles, B.H.; Ward, S.; Todd, J.;

Drobniewski, F.; Wewis, J.; Sawyer, T.; Last, D.; Nichols, C. In Molecular Biology of Microbial Differentiation; Hoch, J.A.; Setlow, P., Eds; American Society for Microbiology: Washington DC, 1985.
3. Burgerjon, A.; Martouret, D. In Microbial Control of Insects and Mites; Burges, H.D.; Hussey, N.W., Eds.; Academic Press: 1971; pp 305-25.
4. Heimpel, A. Ann. Rev. Entomol. 1967, 12, 287-322.
5. Heimpel, A.; Angus, T. J. Insect Pathol. 1959, 1, 152-70.
6. Angus, T. Can. Ent. 1956, 88, 280-83.
7. Dulmage, H.T.; and Cooperators. In Microbial Control of Pests and Plant Diseases 1970-1980; Burges, H.D., Ed.; Academic Press: London; 1981.
8. Krywienczyk, J.; Angus, T. J. Invert. Path. 1967, 9, 126-28.
9. Krywienczyk, J.; Angus, T. J. Invert. Path. 1969, 14, 258-61.
10. Krywienczyk, J.; Dulmage, H.T.; Fast P.G. J. Invert. Path. 1978, 31, 372-75.
11. Krywienczyk, J.; Dulmage, H.T.; Hall, I.M.; Beegle, C.C.; Arakawa, K.Y.; and Fast, P.G. J. Invert. Path. 1981, 37, 62-65.
12. Kronstad, J.W.; Schnepf, H.E.; Whiteley, H.R. J. Bacteriol. 1983, 143, 419-28.
13. Hofte, H.; Whiteley, H.R. Microbiol. Rev. 1989, 53, 242-55.
14. Milne, R. Personal communication.
15. Ge, A.Z.; Shivarova, N.I.; Dean, D.H. Proc. Natl. Acad. Sci. USA 1989, 86, 4037-41.
16. Geiser, M.; Schweitzer, S.; Grimm, C. Gene 1986, 48, 109-18.
17. Hofmann, C., Vanderbruggen, H.; Hofte, H.; Van Rie, J.; Jansens, S.; Van Mellaert, H. Proc. Natl. Acad. Sci. USA 1988, 85, 7844-48.
18. Hofte, H., Van Rie, J.; Jansens, S.; Van Houtven, A.; Vanderbruggen, H.; Vaeck, M. Appl. Environ. Microbiol. 1988, 54, 2010-17.
19. Burges, H.D.; Thomson, E.M.; Latchford, R.A. J. Invert. Pathol. 2 1976, 87-94.
20. Fast, P.G. Can. Ent. 1977, 109, 1515-18.
21. Spies, A.G.; Spence, K.D. Tissue Cell 1985, 17, 379-94.
22. Milne, R.; Fast, P.G. J. Invert. Pathol. 1977, 29, 230-31.
23. Tojo, A.; Aizawa, K. Appl. Environ. Microbiol. 1983, 45, 576-80.
24. Berenbaum, M. Amer. Naturalist 1985, 118, 138-46.
25. Jaquet, F.; Hutter, R.; Luthy, P. Appl. Environ. Microbiol. 1987, 53(3), 500-04.
26. Faust, R., Adams, J.; Heimpel, A. J. Invert. Pathol. 1967, 9, 488-99.
27. Huber, H.; Luthy, P. In Pathogenesis of Invertebrate Microbial Diseases; Davidson, E., Ed.; Allanheld, Osmaun Publishers: New Jersey, 1981; pp. 209-34.
28. Lecadet, M.; Dedonder, R. Bull. Soc. Chim. Biol. 1966, 48, 631-60.
29. Martouret, D. C.r. 11th Congr. Int. Ent. Wien, (1960) 1962, No. 2,849-55.
30. Yamvrias, C. Entomophaga 1962, 7, 101-59.
31. Rivers, D., Vann, C.; Zimmack, H. Proc. Indiana Acad. Sci. 1988 (in press).
32. Lecadet, M.; Martouret, D. J. Invert. Pathol. 1965, 7, 105-08.
33. Chestukhina, G.G.; Kostina, L.I.; Mikhailova, A.L.; Tyurin, S.A.; Klepikova, F.S.; Stepanov, V.M. Arch. Microbiol. 1982, 132, 159-62.
34. Luthy, P. FEMS Microbiol. Lett. 1980, 8, 1-7.
35. Faust, R.M., Hallam, G.M.; Travers, R.S. J. Invert. Pathol. 1974, 24, 365-73.
36. Andrews, R.E.; Bibilos, M.M.; Bulla, L.A., Jr. Appl. Environ. Microbiol. 1985, 50, 737-42.

37. Bulla, L.A., Jr., Kramer, K.J.; Davidson, L.I. J. Bacteriol. 1977, 130, 375-83.
38. Chestukhina, G.G., Zalunin, I.A.; Kostina, L.I.; Kotova, T.S.; Kattrukha, S.P.; Stepanov, V.M. Biochem. J. 1980, 187, 457-65.
39. Bibilos, M.; Andrews, R.E., Jr. Can. J. Microbiol. 1988, 34, 740-47.
40. Applebaum, S.W. In Comprehensive Insect Physiology Biochemistry and Pharmacology; Kerkut, G.A.; Gilbert, L.I., Eds.; Permagon: Toronto, 1985; Vol. 4. p. 279.
41. Eguchi, M.; Kuriyama, K. Comp. Biochem. Physiol. 1983, 76B, 29-34.
42. Houseman, J.G.; Philogene, B.J.R.; Downe, A.E.R. Can. J. Zool. 1989, 67, 864-68.
43. Luthy, P.; Ebersold, H.R. Pharmac. Ther. 1981, 13, 257-83.
44. Tojo, A.; Samasanti, W.; Yoshida, N.; Aizawa, K. Agric. Biol. Chem. 1986, 50, 575-80.
45. Murphy, D. W. Ph.D. Thesis, University of California, 1973.
46. Lecadet, M.-M.; Dedonder, R. Bull. Soc. Chim. Biol. 1966, 48, 661-91.
47. Nagamatsu, Y.; Itai, Y.; Hathnaka, C.; Funatsu, G.; Hayashi, K. Agric. Biol. Chem. 1984, 48, 611-19.
48. Aronson, A. I.; Beckman, W.; Dunn, P. Microbiol. Rev. 1986, 50, 1-24.
49. Bietlot, H.; Carey, P. R.; Choma, C.; Kaplan, H.; Lessard, T.; Pozsgay, M. Biochem. J. 1989, 260, 87-91.
50. Haider, M. Z.; Ellar, D. J. Biochem. J. 1987, 248, 197-201.
51. Hofte, H.; deGreve, H.; Seurinck, J.; Jansens, S.; Mahillon, J.; Ampe, C.; Vandekerckhove, J.; Vandergruggen, H.; van Montagu, M.; Zabeau, M.; Vaek, M. Eur. J. Biochem. 1986, 161, 278-80.
52. Hofmann, C.; Luthy, P.; Hutter, R.; Pliska, V. Eur. J. Biochem. 1988, 156, 531-40.
53. VanRie, J.; Jansens, S.; Hofte, H.; Degheele, D.; VanMellaert, H. Eur. J. Biochem. 1989, (in press)
54. Wolfersberger, M. Experimentia 1990, (in press)

RECEIVED February 22, 1990

Chapter 5

In Vitro Analyses of *Bacillus thuringiensis* δ-Endotoxin Action

George E. Schwab and Paul Culver

Department of Biochemistry, Mycogen Corporation, 5451 Oberlin Drive, San Diego, CA 92121

The purpose of the present report is to review the literature as it pertains to the mechanism of action of Bacillus thuringiensis (B.t.) endotoxins as determined by in vitro experimentation. As a means toward providing a focus to this report, we will only discuss those studies that employ lepidopteran-derived model systems and lepidopteran-specific B.t. endotoxins.

The use of cell culture and isolated tissue preparations, which comprise the primary model systems for in vitro experimentation, has certain limitations. In most instances, the insect cells used in tissue culture are not considered the primary target for endotoxin interactions, ie. the columnar cells of the midgut epithelium. This is due to the fact that attempts at establishing a continuous line of insect derived midgut epithelial cells have not been successful. In like manner, the use of isolated target organ tissues like the midgut, or tissues or subfractions prepared therefrom, may destroy effective cellular interactions. Finally, most in vitro approaches rely upon a prerequisite activation of the protoxin to an active toxin (usually by solubilization and/or limited proteolysis) prior to its introduction to the in vitro system (1,2). In some instances this is achieved by using trypsin. Although trypsin has been shown to mimic the action of several lepidopteran gut proteases, care should be used in the interpretation of results obtained by these means. The gut proteases of several lepidopteran strains are capable of differentially processing B.t. endotoxins and the correct scission of the B.t. endotoxin may be a chief determinant in specificity (3). Despite these and other limitations, basic information regarding the mechanisms of B.t. endotoxin-mediated insect toxication continues to be provided. It is doubtful that in vitro methods of analysis will ever afford a viable alternative to in vivo methods for the quantitation of toxin bioactivity.

Endotoxin Pathogenesis

Three decades ago, experiments were undertaken to assess B.t. toxin pathogenesis. These were largely performed using histopathologic

0097–6156/90/0432–0036$06.00/0

and histochemical techniques on tissues derived from B.t. endotoxin-treated larvae. As a result of these studies, the midgut was found to be the most susceptible organ (4). Among the physiologic changes observed in the midgut following administration of the B.t. endotoxin were: swelling of the microvilli lining the lumen of the midgut epithelia; vacuole formation in the cytoplasm; separation of the epithelial cells from the lumen; degeneration of the peritrophic membrane; and morphologic changes in many cellular organelles (ribosomes, mitochondria, rough endoplasmic reticulum, etc.)(4-12). The chronology of the above events favors the columnar cells of the midgut epithelia as the earliest cellular site of involvement of B.t. endotoxin-mediated activity (cf. 13).

A departure from the cataloguing of morphologic anomalies associated with B.t. endotoxin activity by light and EM microscopy was provided by Heimpel and Angus (4). These authors measured an increase in the pH of the hemolymph of B. mori following disruption of the midgut epithelium. This led to the suggestion that the actual toxic event caused by the endotoxin was through a disruption of the epithelium that resulted in an ionic imbalance. Once the epithelium is compromised, the alkaline midgut environment, rich in potassium ions, is allowed unrestricted passage into the hemolymph. This results in an increase in the hemolymph pH and subsequent paralysis of the organism. Nishiitsuji-Uwo and Endo (6,7) extended these observations to the larva of P. rapae, L. dispar, E. cautella and G. mellonella. Following addition of toxin to these insects, histopathologic symptoms similar to those observed in B. mori were noted. Furthermore, an increase in hemolymph pH (P. rapae) and potassium ion concentration in the hemolymph (P. rapae, L. dispar, E. cautella and G. mellonella) was observed.

Thus a mechanistic scenario for B.t. toxication began to evolve. Somehow, the B.t. endotoxin is able to recognize and discriminate among various cell types in the insect gut. Following such recognition, the affected cells swell and eventually burst, the result being a disproportionate amount of potassium ion entering the hemocoel. The details describing the molecular basis of these events associated with B.t. endotoxin-mediated action on the whole organism form the framework for the following in vitro approaches.

Isolated Insect Midguts

The use of isolated midguts permits a model system with which to investigate the biochemistry of the toxication process in more detail separate from the complexities presented by the intact organism.

Using isolated midgut of M. sexta, Harvey and Wolfersberger (14) measured both the electrophysiological and biochemical parameters affecting ion transport. B.t. endotoxin-mediated inhibition of potassium transport was proposed to arise by interference with the efflux of this electrolyte from the lumen to the hemocoel. Griego et al. (15) in a similar study, monitored both the short circuit current (SSC) and transepithelial transport of potassium ions in B.t. endotoxin-treated midgut of M. sexta. Each of the two parameters was shown to be inhibited one hour following ingestion. Utilizing the same type of midgut preparation from M. sexta, Gupta et al. (16) presented evidence that B.t. kurstaki disrupts potassium ion tranport across the goblet cell by an interaction at the apical membrane. This in turn leads to destruction of the potassium ion gradient as well as the potential difference across the membrane. Subsequent to such disruption in the equilibrium provided by this gating mechanism would be a rise in the intracellular pH. By employing the divalent cations, barium and calcium, Crawford and Harvey (17) demonstrated a reversal of the B.t. endotoxin-induced inhibition of the short circuit current in M. sexta midgut preparations. This in turn provided additional evidence for an endotoxin effect on transmembrane ion flux.

Wolfersberger and Spaeth (18) determined whether a correlation could be made between the in vitro and in vivo toxicities of various B.t. endotoxins. Using midgut from M. sexta, the authors examined a panel of twenty-two spore-endotoxin preparations with respect to the inhibition of net potassium transport in isolated midgut preparations. These effects were then compared to the larvicidal activity of these spore-endotoxin preparations in vivo. They found that the in vitro assay was a good predictor of toxicity.

A departure from the classic isolated midgut preparation was designed by Yunovitz et al. (19). These authors dissected the midgut from fifth instar S. littoralis. A flat-sheet preparation was cut from the midgut (both in the presence and absence of the peritrophic membrane) and mounted in a perfusion chamber. Ringer solutions of pH 10 and 6.5 were used to bathe the lumen and hemocoel sides, respectively. The effects of a highly purified protein fraction obtained from B.t. entomocidus on this system were gauged by measuring the activity of reduced glutathione S-transferase released into the lumen media as a result of endotoxin-mediated rupture of the endothelial cells. They found that the effects of the toxin were mitigated when the peritrophic membrane was present as a result of toxin aggregate formation.

Studies utilizing isolated tissues have not been confined solely to midgut preparations. Reisner et al. (20) examined the effect of B.t. kurstaki endotoxin (HD-73 in both the protoxin and activated toxin configurations) on isolated malpighian tubules of C. ethlius larva. Both inhibition of urine secretion and massive cytolysis were observed after direct introduction of the activated toxin into the lumen. The pathogenesis of toxin activity on these tissues mirror that observed in the midgut. This study serves to demonstrate that the effects of the toxin are not confined only to the epithelia of the midgut.

Insect Cell Lines

Over the past two decades a number of stable cell lines have been established from a variety of lepidopteran species. Several of these lines (Table I) have been evaluated for their capacity to serve as in vitro models for the analysis of endotoxin action and in many cases, have served this function well. Several distinct advantages accrue to the use of cell lines in this regard. Owing to the relative simplicity of the model system as compared to insect bioassay, it becomes feasible to focus on the primary molecular events leading to toxicity. As a bioassay tool, the cellular assay could be run more frequently with less variability in results than with the whole insect. Probably of greatest potential value is the use of cultured cell systems for conducting parallel studies of receptor occupation by endotoxin and the cytotoxic response.

Table I. Lepidopteran Cell Lines

Line	Insect Source	Common Name	Tissue Source
CF-1	Choristoneura fumiferana	Spruce budworm	Whole larvae
HV-AM1	Heliothis virescens	Tobacco budworm	Pupal ovary
HZ-1075	Heliothis zea	Cotton bollworm	Adult ovary
LD-652Y	Lymantria dispar	Gypsy moth	---
MD-108	Malacosoma disstria	Tent caterpillar	Larval hemocytes
---	Mamestra brassicae	Cabbage moth	---
CHE	Manduca sexta	Tobacco hornworm	Embryonic
SF-21AE	Spodoptera frugiperda	Fall armyworm	Pupal ovary
TN-368	Trichoplusia ni	Cabbage looper	Adult ovary

However, certain disadvantages are encountered in the application of insect cell lines to the study of endotoxins. As was discussed previously, insect cell assays are conducted in an environment significantly different from the insect midgut, involve cells derived from non-midgut sources, and necessitate prior activation of protoxins by enzymic digestion. Obviously there are ways to circumvent most all of these disadvantages. In vitro processing of protoxins with purified enzymes or insect gut juices is routinely done. Assay pH is elevated as high as possible even though full toxin solubilization cannot be achieved. Pluripotentiality of some cell lines might obviate doubts concerning the relevance of findings obtained from cultured cells of non-midgut origin. Essentially, the broader relevance of information gleaned from cultured cell systems is often challenged, but this rarely impacts on the ultimate utility of the in vitro approach.

Since it has long been known that the morphological effects of endotoxin action in the insect center on swelling and lysis of cells of the midgut epithelium, in vitro characterizations of toxin action with insect cell lines began with morphological analysis. The earliest published work in this area is that of Murphy et al. (2), which demonstrated generalized swelling of several lepidopteran cell lines in response to incubation with purified toxin crystals from B.t. kurstaki which had been enzymatically processed with gut proteases from T. ni. Cytotoxicity was completely abolished by heat treatment of the toxin, or by prior incubation of the toxin with specific antiserum, thereby confirming the protein identity of the toxic component. Since this report, other investigators have demonstrated in vitro cytotoxicity with B.t. kurstaki (21-26) and B.t. aizawai endotoxins (24,27-30) in a variety of insect cell lines. In addition, a broad spectrum of B.t. toxins has been screened against CF-1 and CHE cells for cytotoxic activity (31). For the most part, the index of cytotoxicity has been qualitative, i.e. observations of cell swelling or lysis or the capacity to exclude vital dyes such as trypan blue. However, quantitative methods have been introduced such as the determination of cellular ATP by the luciferin/luciferase method (23), uptake of rubidium-86 (21), or measurement of intracellular sodium and potassium (29) as a function of exposure to endotoxin.

With respect to the mode of action of endotoxins, a few candidate mechanisms have emerged from studies with insect cell lines. An early report by Nishiitsutsuji-Uwo et al. (27) described a dependence of endotoxin cytotoxicity on sodium ion concentration. By varying sodium chloride concentration while maintaining constant osmolarity with sucrose, it was demonstrated that endotoxin cytotoxicity requires a threshold NaCl concentration of 100 - 150 mM. Further evidence of a relationship between endotoxin cytopathy and ion fluxes exists in the observation of increasing sodium ion and decreasing potassium ion concentrations inside cells undergoing toxin-induced swelling (29). Furthermore, this swelling was blocked by prior treatment with the sodium channel blocker tetrodotoxin, and was potentiated by the potassium channel blocker 4-aminopyridine. Thus it appears that the application of toxin to cells generates an increased permeability to ions accompanied by an influx of water and cellular swelling. This mechanism for endotoxin action has been further substantiated and expanded by experiments which demonstrated protective effects of externally applied solutes as well as leakage of internal radioactive markers in response to toxin application (32). These latter findings were consistent with an ionophore mechanism for endotoxins wherein the toxin induces pores in the plasma membrane through which ion flux occurs. The net effect of this is an influx of water and subsequent cellular swelling and lysis. David Ellar's group refers to this mechanism as colloidal osmosis. Examination of the osmotic protection afforded by various size solutes led them to the assignment of a pore radius in the range of 0.5 - 1.0 nm for B.t. kurstaki (32) and B.t. aizawai (32,33).

Additional cellular effects have been reported for endotoxins. A K+,H+-ATPase activity in CHE cells has been shown by English and Cantley (21) to be inhibited by toxin. Interestingly, the locus of this ATPase is such that there is an implied translocation of B.t. toxin from the lumen surface of the midgut cell to the hemocōel in order to exert its inhibitory effect. This would represent a significant departure from the more widely held hypothesis of an action limited to pore formation at the lumen surface of the midgut epithelial cell. Additional evidence of an ATPase involvement comes from demonstrations that cytotoxicity is ameliorated by ouabain, a specific inhibitor of Na+,K+-ATPase (29). Finally, one report by Knowles and Farndale (24) linked increases in intracellular cAMP levels to endotoxin application, though the authors concluded that this was a secondary effect and neither necessary nor sufficient for cell lysis.

It has long been presumed that the marked species specificity observed for endotoxins implies the existence of toxin specific receptors in affected insect species. Studies with insect cell lines have begun to yield important information regarding the nature of the toxin receptor. Knowles et al. used the CF-1 cell line to determine whether various biological compounds would interfere with endotoxin cytolysis of cells (34). Lipids derived from various sources did not attenuate toxin action, whereas the amino sugars, N-acetylgalactosamine and N-acetylneuraminic acid, did. Agglutinins from wheat germ and soy bean also afforded protection from cytotoxicity. Thus it was concluded that B.t. endotoxins exhibit lectin-like binding to a surface receptor that is probably a glycoprotein.

In a direct approach to receptor identification, CF-1 cell proteins separated by SDS-polyacrylamide gel electrophoresis were transferred to nitrocellulose and probed with either radioiodinated toxin from B.t. kurstaki or radioiodinated soy bean agglutinin (25). Toxin and agglutinin reacted with identical but multiple protein components. CF-1 cells were covalently surface labeled with tritiated sodium borohydride to determine which protein components were exposed at the cell surface and might be the putative receptor(s). Reasoning that a plasma membrane receptor would likely bind toxin at the exterior, the authors were able to identify a single 146 kDa protein which was both surface-labeled and bound B.t. kurstaki toxin. A similar direct binding determination was used by Haider and Ellar (30) to identify a putative receptor for the endotoxin of B.t. aizawai. By probing blots of membrane proteins from CF-1 and S. frugiperda cells with purified, activated toxin, followed by reaction with a secondary antibody-peroxidase conjugate, a 68 kDa binding protein was identified. Probing western blots prepared from CF-1 cell proteins with iodinated B.t. aizawai toxin also revealed a 68 kDa binding protein and an additional 120 kDa putative receptor (33). Though the approach of probing blots with ligand does identify a binding protein, it is unclear whether the binding protein represents the true receptor. Confirmation would require receptor purification and its reconstitution into a system in which the receptor's functional response to toxin application could be demonstrated.

McCarthy et al. (22) present evidence that endotoxins and Baculoviruses might share affinity for the same receptor. When the extracellular virus of Autographa californica was incubated with increasing concentrations of endotoxin and subsequently assayed on T. ni cells, viral infectivity in these cells was reduced. The authors suggest that extracellular virus gains intracellular access after a primary surface recognition and binding event involving the endotoxin receptor, and that toxin competitively inhibits binding of virus to the receptor. The possibility of shared receptor recognition regions in viral and toxin protein structure might have some bearing on evolutionary questions regarding the origin of endotoxins.

Brush Border Membrane Vesicles

As stated in the preceding sections, the epithelium of the insect midgut is a complex multicompartmental system. It is the site of intracellular and transcellular transport function. Both ionic regulation and mineral accumulation are effected by the cells that comprise this tissue. In an attempt to gain a deeper understanding of the effects of B.t. endotoxin on the midgut epithelium, lumen plasma membranes of the lumen microvilli, more commonly referred to as brush border membrane vesicles (BBMV), have been employed as a model for analysis of the molecular basis of the toxication process.

Owing to the relative rigidity of the brush border membrane, once the midgut cell preparation is homogenized, the enrichment of the BBMV by differential centrifugation is easily accomplished. In addition, by exploiting the inherent charge properties of the lumen membrane, magnesium or calcium driven precipitation of the brush border membrane provides a facile means to achieve the preparation of these vesicles. Several marker enzymes can be monitored to assess the enrichment of the BBMV. Another attractive characteristic of BBMV preparations is the preservation of the sidedness of the vesicles. Specifically, the orientation of the membrane is as normally encountered in the intact midgut epithelium. In addition, the buffer components used in these investigations can be carefully controlled to suit the particular study. Nevertheless, there are drawbacks to the use of BBMV as a model. The vesicles are an artifact of the procedures used to make them such that alterations in permeability and/or function of the vesicles may result. Overall, the data obtained from experiments conducted with BBMV are usually considered to be qualitative in nature rather than quantitative.

Using vesicles prepared from B. mori, Tojo (35) observed gross morphological alterations of vesicles when incubated with full length B.t. kurstaki protoxin. When the protoxin was activated to a limit peptide by B. mori gut proteases, the denaturation of the vesicles was increased. Studies using BBMV prepared from the midguts of P. cynthia and B. mori resulted in the characterization of a potassium-dependent co-transport mechanism for several hydrophilic amino acids (36-42). Sacchi and co-workers (43) measured amino acid transport in BBMV prepared from fifth instar P. brassicae midgut. They found that trypsin-activated endotoxins obtained from B.t. kurstaki and B.t. thuringiensis inhibited the uptake of several neutral and basic amino acids. In experiments in which the potassium concentration was equilibrated on both sides of the membrane, the endotoxin lost its ability to inhibit the uptake of amino acids. This indicated that the toxin specifically affects the potassium ion permeability of the vesicle instaed of acting directly on amino acid potassium co-transport.

The use of BBMV has not been limited to electrophysical and chemicophysical measurements. Several groups have investigated the binding of B.t. endotoxins to surface receptors on BBMV. Hofmann, et al. (44) reported on the presence of a putative B.t. endotoxin receptor on the surface of BBMV prepared from P. brassicae. Although initial experiments conducted with radioiodinated endotoxin gave evidence of high and low affinity receptor populations, subsequent experimentation with a non-iodinated toxin resulted in a single binding-site population with an intermediate dissociation constant. This discrepancy is believed due to the apparent augmentation in hydrophobicity imparted to the toxin by iodination. Further experimentation compared the specificity of binding among different BBMV preparations obtained from P. brassicae and M. sexta (45). In this study, toxicity of a particular endotoxin was found to correlate with binding. In addition, evidence for the existence of discrete binding sites was found for heterologous B.t. endotoxins.

Conclusion

While a broad consensus on the actual mechanism of B.t. toxicity does not yet exist, in surveying the current in vitro studies it seems that the mechanism summarized in Ellar's colloidal osmotic hypothesis (32) is most consistent with the current data. The proposed mechanism is a two step process in which toxin first recognizes and binds to a surface receptor and secondly, induces the formation of pores in the membrane. Such disruption results in a bidirectional ion flux, the net result being an influx of water leading to cellular swelling and bursting. Clearly this working hypothesis is one that is testable from a variety of approaches and would be very useful as a framework for future experimental design.

A recurring problem that surfaces in research addressing the toxication mechanism of B.t. endotoxin relates to ambiguities in the identity of the toxins used. Differing methods of toxin purification, extent of spore contamination and enzymatic processing can give rise to a diversity of protein species, and it is very likely that what is identified as a proteolytically activated toxin from B.t. kurstaki can vary considerably from one laboratory to another. This is an acute problem when data pertaining to the electrophoretic purity of the proteins used is not presented.

Despite the deficiencies and shortcomings noted above, it is certain that continued progress will be made in establishing the mechanism for B.t. entomotoxicity and that in vitro methods of analysis will play a critical role in this process.

Literature Cited

1. Cooksey, K.E. In: Microbial Control of Insects and Mites; Academic Press: New York, NY, USA, 1971; 247-274.
2. Murphy, D.W.; Sohi, S.S.; Fast, P.G. Science 1976, 194, 954-956.
3. Haider, M.Z.; Knowles, B.H.; Ellar, D.J. Eur. J. Biochem. 1986, 156, 531-540.
4. Heimpel, A.M.; Angus, T.A. J. Insect Pathol. 1959, 1, 152-170.
5. Heimpel, A.M.; Angus, T.A. Bacterial. Rev. 1960, 24, 266-288.
6. Nishiitsutsuji-Uwo, J.; Endo, Y. Appl. Entomol. Zool. 1981, 16, 79-87.
7. Nishiitsutsuji-Uwo, J.; Endo, Y. Appl. Entomol. Zool. 1981, 16, 225-230.
8. Mathavan, S.; Sudha, P.M.; Pechimuthu, S.M. J. Invertebr. Pathol. 1989, 53, 217-227.
9. Endo, Y.; Nishiitsutsuji-Uwo, J. J. Invertebr. Pathol. 1980, 36, 90-103.
10. Ebersold, H.R.; Luethy, P.; Mueller, M. Mitt. Schweitz Entomol. Ges. 1977, 50, 269-276.
11. Sutter, G.R.; Raun, E.S. J. Invertebr. Pathol. 1967, 9, 90-103.
12. Percy, J.; Fast, P.G. J. Invertebr. Pathol. 1983, 41, 86-98.
13. Luethy, P.; Ebersold, H.R. In: Pathogenesis of Invertebrate Microbial Diseases; Allenheld Osmum, Totowa, NJ, USA; 235-267.
14. Harvey, W.R.; Wolfersberger, M.G. J. Exp. Biol. 1979, 83, 293-304.
15. Griego, V.M.; Moffet, D.; Spence, K.D. J. Insect Physiol. 1979, 25, 283-288.
16. Gupta, B.L.; Dow, J.A.T.; Hall, T.A.; Harvey, W.R. J. Cell Sci. 1985, 74, 137-154.
17. Crawford, D.N.; Harvey, W.R. J. Exp. Biol. 1988, 137, 277-286.
18. Wolfersberger, M.G.; Spaeth, D.D. J. Appl. Entomol. 1987, 103, 138-141.
19. Yunovitz, H.; Sneh, B.; Schuster, S.; Oron, U.; Broza, M.; Yawetz, A.J. J. Invertebr. Pathol. 1986, 48, 223-231.
20. Reisner, W.M.; Feir, D.J.; Lavrik, P.B.; Ryerse, J.S. J. Invertebr. Pathol. 1989, 54, 175-190.
21. English, L.H.; Cantley, L.C. J. Membrane Biol. 1985, 85, 199-205.
22. McCarthy, W.J.; Aronson, J.N.; Labenberg, J. In vitro Cellular and Dev. Biol. 1988, 24, 59-64.
23. Johnson, D.E. J. Invertebr. Pathol. 1981, 38, 94-101.
24. Knowles, B.H.; Farndale, R.W. Biochem. J. 1988, 253, 235-241.
25. Knowles, B.H.; Ellar, D.J. J. Cell Sci. 1986, 83, 89-101.
26. Knowles, B.H.; Thomas, W.E.; Ellar, D.J. FEBS Lett. 1984, 168, 197-202.
27. Nishiitsutsuji-Uwo, J.; Endo, Y.; Himeno, M.J. Invertebr. Pathol. 1979, 34, 267-275.
28. Himeno, M. In: Biotechnology in Invertebrate Pathology and Cell Culture; Academic Press: San Diego, CA, USA, 1987; Chapter 3.
29. Himeno, M.; Koyama, N.; Funato, T.; Komano, T. Agric. Biol. Chem. 1985, 49, 1461-1468.
30. Haider, M.Z.; Ellar, D.J. Mol. Microbiol. 1987, 1, 59-66.
31. Johnson, D.E. FEMS Microbiol. Lett. 1987, 43, 121-125.
32. Knowles, B.H.; Ellar, D.J. Biochim. Biophys. Acta 1987, 924, 509-518.

33. Haider, M.J.; Ellar, D.J. Biochem. J. 1987, 248, 197-201.
34. English, L.H.; Cantley, L.C. J. Biol. Chem. 1986, 261, 1170-1173.
35. Tojo, A. Appl. Environ. Microbiol. 1986, 51, 630-633.
36. Wolfsberger, M.; Luethy, P.; Maurer, A.; Parenti, P.; Sacchi, F.V.; Giordana, B.; Hanozet, G.M. Comp. Biochem. Physiol. 1987, 86A, 301-308.
37. Hanozet, G.M.; Giordana, B.; Parenti, P.; Guerritore, A.J. Membrane Biol. 1984, 81, 233-240.
38. Sacchi, V.F.; Hanozet, G.M.; Giordana, B. J. Exp. Biol. 1984, 108, 329-339.
39. Hanozet, G.M.; Giordana, B.; Sacchi, V.F. Biochim. Biophys. Acta 1980, 596, 481-486.
40. Giordana, B.; Sacchi, V.F.; Hanozet, G.M. Biochim. Biophys. Acta 1982, 692, 81-88.
41. Giordana, B.; Parenti, P.; Hanozet, G.M. J. Membrane Biol. 1985, 88, 45-53
42. Parenti, P.; Giordana, B.; Sacchi, V.F.; Hanozet, G.M.; Guerritore, A. J. Exp. Biol. 1985, 116, 69-78.
43. Sacchi, F.V.; Parenti, P.; Hanozet, G.M.; Giordana, B.; Luthy, P.; Wolfersberger, M.G. FEBS Lett. 1986, 204, 213-218.
44. Hofmann, C.; Luthy, P.; Hutter, R.; Pliska, V. Eur. J. Biochem. 1988, 173, 85-92.
45. Hoffman, C.; Vanderbruggen, H.; Hofte, H.; Van Rie, J.;Jansens, S.; Van Mellaert, H. Proc. Nat. Acad. Sci. USA 1988, 85, 7844-7848.

RECEIVED March 1, 1990

Chapter 6

Identification of Entomocidal Toxins of *Bacillus thuringiensis* by High-Performance Liquid Chromatography

Takashi Yamamoto

Sandoz Crop Protection Corporation, 975 California Avenue, Palo Alto, CA 94304–1104

In this review, the methodology developed to characterize various Bacillus thuringiensis (B.t.) crystal proteins by high performance liquid chromatography (HPLC) is described. B.t. crystal protein has been highly purified by column chromatography to be peptide-mapped by trypsin digestion. Peptides in the mixture are separated by HPLC, and the separation is chromatographically recorded. The chromatogram obtained is very specific to the protein being analyzed and serves as a finger print. Peptide mapping may be used to determine the expression level of the crystal gene. The peptide mapping, when combined with high resolution liquid matrix secondary ion mass spectrometry, can confirm the amino acid sequence of the crystal protein. Purification methods of the crystal protein, including size-exclusion and affinity column chromatography, are also reviewed.

Background

Bacillus thuringiensis (B.t.) is known to be pathogenic to a variety of insects, notably to lepidopteran species. Numerous strains have been isolated from dead insects, soil samples, grain dusts, etc., and have been classified into some 20 subspecies, primarily on the basis of flagellar antigens (1). B.t. grows in an ordinary culture medium (e.g. nutrient broth) at 30°C with vigorous aeration. When B.t. exhausts nutrients which support its growth, the bacterium produces a spore and one or more insecticidal proteins in crystalline form. After the sporulation, cells are eventually lysed and release free spores and crystals into the culture medium.

Normally, larvae of insects susceptible to B.t. have alkaline digestive juice whose pH is as high as 11. When an insect larva ingests B.t. crystals, the crystals are dissociated in the digestive juice. Most B.t. crystals toxic to

0097–6156/90/0432–0046$06.00/0

lepidopteran species are made of proteins having molecular weights around 130,000 (2). The 130-kDa crystal protein liberated from the crystal matrix is then digested by proteinases in the digestive juice into roughly one half of its original size (3). It has been shown that the N-terminal half of the protein is relatively resistant to proteinases and contains the full insecticidal activity (4).

B.t. kurstaki HD-1, isolated as a high potency strain (5), has been commercially produced for many years in the U.S. HD-1 produces two kinds of crystals in one cell, a bipyramidal crystal made of 130-kDa proteins and a cuboidal crystal of 65-kDa protein or proteins (6). Hall et al. (7) found that HD-1 shows significant toxicity to mosquito larvae, while HD-73, a different strain of subsp. kurstaki, does not. The mosquitocidal activity of HD-1 is attributed to the 65-kDa protein termed P2 which does not occur in HD-73 (8). P2 has been found in many other kurstaki strains such as HD-263 (9).

Gonzalez et al. (10) have shown that B.t. kurstaki strains harbor a variety of plasmids, some of which have been implicated in crystal production. Kronstad et al.(11) indicated that crystal genes (cry) encoding the 130-kDa protein are located on several different plasmids of HD-1. Schnepf and Whiteley (12) cloned the first cry gene in E. coli from an HD-1 strain called "HD1-Dipel." Subsequently, several additional cry genes encoding the 130-kDa protein were cloned from HD-1 and other kurstaki strains (13-17). These cry genes were classified into three types as cryIA(a), cryIA(b), and cryIA(c) (18).

This paper details applications of HPLC developed to characterize the crystal protein demonstrating that HPLC can be used to determine expression levels of cryIA-type genes.

Isolation and Solubilization of B.t. Crystals

B.t. does not require a special medium for its growth. Nonetheless, we have found that a medium containing peptonized-milk nutrient or casein hydrolyzate along with potassium phosphate supports a better growth and crystal production (T. Yamamoto and C. C. Beegle, unpublished). Table I lists the components of a medium called "CYS" which we have developed for the crystal isolation. Due to the high concentration of nutrients in this

Table I. CYS medium

Bacto-casitone	10	g/l
Bacto-yeast extract	2	g/l
Glucose	5	g/l
KH_2PO_4	1	g/l
$MgCl_2 \cdot 6H_2O$	0.5	mM
$MnCl_2 \cdot 4H_2O$	0.05	mM
$ZnSO_4 \cdot 7H_2O$	0.05	mM
$FeCl_3 \cdot 6H_2O$	0.05	mM
$CaCl_2 \cdot 2H_2O$	0.2	mM

Adjust pH to pH 7 with 2 N NaOH.

medium, it supports a relatively high cell density at 30°C. However, in this medium B.t. grows slowly, requiring over 48 hr to complete sporulation. The B.t. crystal has been isolated in high purity by density-gradient centrifugation. Since the crystal protein is subsequently purified by column chromatography, the purity of the crystal is not critical for peptide mapping of the protein. However, it is essential to remove endogenous proteinases from the crystal preparation to prevent further nonspecific digestion of the protein. NaCl has been used to facilitate the removal of the proteinases. Angus (19) treated the crystal with 1 M NaCl at 37°C; we found 0.5 M NaCl gave better proteinase removal (20). It is important to process the crystal with NaCl at a warm temperature (e.g. 37°C) to break hydrophobic interaction of the protein with contaminating proteinases. Treatment of the crystal with 0.5 M NaCl must be repeated until the presence of proteinases is no longer detected when the crystal is solubilized. Care must be taken not to disrupt any vegetative B.t. cells, if present, as they release additional amounts of proteinase. Alternatively, vegetative cells should be removed before the NaCl treatment.

The crystal is solubilized at pH 12. A pH as low as 10 can be used if a disulfide-reducing agent, such as 2-mercaptoethanol or dithiothreitol (DTT) is added to the solubilizing medium (21).

Isolation of the Crystal Protein

The components of the B.t. crystal can be analyzed by sodium dodecyl sulfate polyacrylamide gel electrophoresis (SDS-PAGE). SDS-PAGE has shown that crystals of many B.t. strains are composed of 130 to 135-kDa proteins (19). Yamamoto and MacLaughlin (8) found that the crystal of HD-1 contains an additional protein of 65-kDa. The 65-kDa protein, termed P2, is responsible for HD-1's mosquitocidal activity and has been isolated from the 135-kDa protein (called P1) by Sephacryl column chromatography. Figure 1 shows a typical separation of components of the HD-1 crystal (8). In this experiment, the HD-1 crystal is solubilized in 2% mercaptoethanol and NaOH at pH 10 and centrifuged to remove any insoluble material. If the crystal preparation contains spores, they remain insoluble under these conditions. The supernatant is applied to a Sephacryl S-300 column without neutralization, and the crystal protein is eluted with 50 mM Tris·HCl, pH 8. Sephacryl S-300 separates proteins by size. P1 elutes slightly after the void volume. The elution volume of P2 is much larger than that expected from its size. This delay in elution of P2 indicates that the protein has a weak affinity to Sephacryl. To take advantage of the affinity of P2 with the column matrix, proteins were eluted from the column with a low flow speed (i.e. 5 ml/cm^{2}hr).

SDS-PAGE and fused-rocket immunoelectrophoresis were employed to assess the purity of P1 and P2. Immunoelectrophoresis is particularly useful for identifying fractions containing highly purified P1 and P2. As many as 40 fractions can be analyzed on one electrophoresis plate (8). SDS-PAGE of P1 and P2 isolated by Sephacryl column chromatography is shown in Figure 2. No significant amount of impurity was found. The isoelectric points of P1 and P2 are pH 4.4 and pH 10.7, respectively (8). Since the solubility of P1 at the isoelectric point is very low, almost all of the P1 in the solution is precipitated at pH 4.4. Furthermore, the acid-precipitated P1 is easily solubilized in 10 mM Tris·HCl, pH 8. This permits a rapid and easy method to concentrate

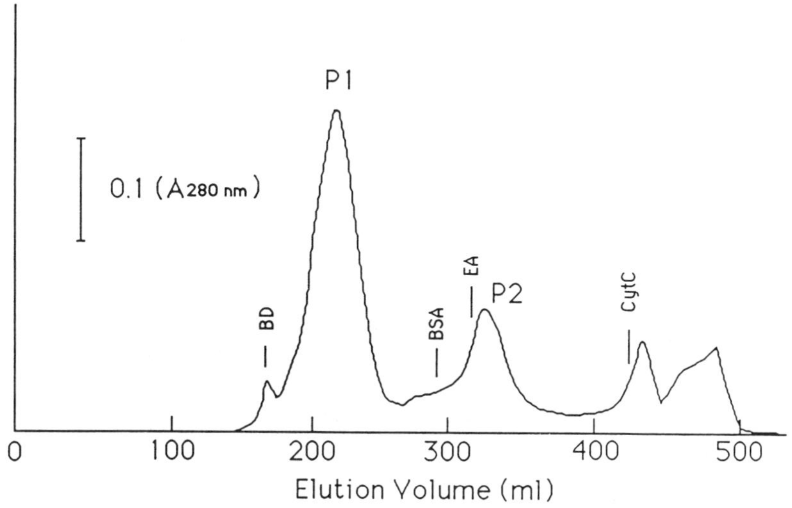

Figure 1. Isolation of the crystal protein by Sephacryl S-300 column chromatography. The B.t. crystal was solubilized in 2% 2-mercaptoethanol at pH 10.7 titrated with NaOH. Elution was performed with 50 mM Tris·HCl, pH 8, containing 0.1% 2-mercaptoethanol and 1 mM EDTA and monitored at 280 nm. The column (2.6 X 100 cm) was calibrated with BD, blue dextran; BSA, bovine serum albumin; EA, egg albumin; CytC, cytochrome C. Molecular weights were determined with myosin (200,000), ß-galactosidase (116,000), phosphorylase b (94,000), bovine serum albumin (68,000), and egg albumin (43,000). (Reproduced from ref. 8 by kind permission from the Academic Press.)

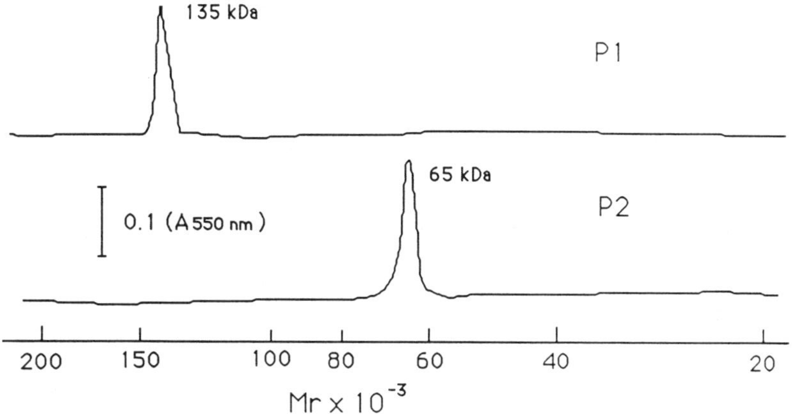

Figure 2. SDS-PAGE scan of P1 and P2 isolated by Sephacryl column chromatography. P1 and P2 were isolated as shown in Fig. 1 and treated with 2% SDS at 100°C. SDS-PAGE was performed in 10% gel, stained with Coomassie blue, and scanned by a densitometer. (Reproduced from ref. 8 by kind permission from the Academic Press.)

P1. If there is any difficulty in solubilizing the acid-precipitated P1, P1 could be partly digested by contaminating proteinases. The C-terminal portion of the P1 molecule is quite sensitive to proteinase digestion. Partially digested P1 has very low solubility at a neutral or slightly alkaline pH.

HPLC Peptide Mapping

Computerization of HPLC allows separation of a complex peptide mixture with high precision and reproducibility (22). It has been demonstrated that two peptides that differ by a single amino acid residue can be separated by HPLC (23). This technique can be used to produce a finger print of the B.t. crystal protein. A typical peptide map of P1 is shown in Figure 3. It was produced by trypsin digestion of purified P1, using a trypsin treated with TPCK (L-1-tosylamide-2-phenylethyl chloromethyl ketone) that inhibits contaminating chymotrypsin. P1 isolated by Sephacryl column chromatography was dissolved in 100 mM Tris·HCl, pH 8, at 10 mg/ml. Digestion was carried out at a ratio of 1:100 (protein: TPCK-treated trypsin) at 37°C for 4 hr. The same amount of trypsin was added at 2 hr during the incubation to ensure complete digestion. After the digestion, 100 μl of the reaction mixture were mixed with 25 μl glacial acetic acid. Under this condition, no precipitation should appear. If any insoluble matter is observed, it may be an indication that the digestion is not completed. The mixture was filtered through a 0.45 μ filter and 10 μl of the filtrate were injected onto the HPLC.

To separate peptides by HPLC, a C18 reversed-phase column was used. The elution was made first with 0.1% phosphoric acid for 0.5 min, followed by gradually increasing acetonitrile from 0% to 30% over a 50-min period, and then to 50% in 5 min. The separation of peptides was monitored at 215 nm.

Peptide Mapping of P1 Isolated from Different Serotypes

To compare P1s produced by different serotypes of B.t., over 20 strains were chosen for HPLC peptide mapping (24). Figure 4 shows HPLC peptide maps of P1s from these B.t. strains. Comparison of these maps revealed a high degree of similarity between two different serotypes. As shown in Figure 4, the peptide map of HD-14 (subsp. *thuringiensis*) is almost identical to that of HD-11 (*aizawai*). According to USDA bioassay data (H. T. Dulmage, unpublished), crystal preparations from these two strains are very similar in activity spectrum. Both crystals are highly active against the silkworm (*Bombyx mori*) but are not so active against the cabbage looper (*Trichoplusia ni*) and tobacco budworm (*Heliothis virescens*). Figure 4 also reveals strong heterogeneities between two strains from one serotype. They are different in activity when assayed in several different insect species (H. T. Dulmage, unpublished).

It is now believed that many B.t. strains produce several P1s in one cell (15). It is possible that maps for some of those strains in Figure 4 are derived from mixtures of several distinct proteins. Nevertheless, the peptide mapping of P1 described above produces highly specific finger prints and can be used to identify B.t. crystal proteins.

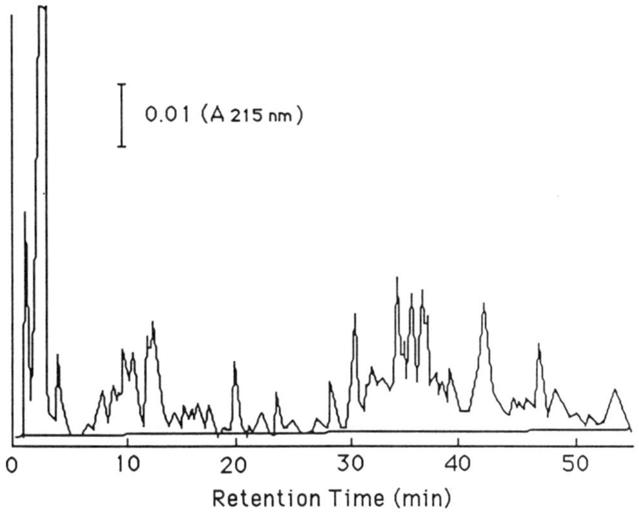

Figure 3. HPLC peptide map of P1 isolated from B.t. kurstaki HD-254 by Sephacryl column chromatography. The isolated P1 was digested with trypsin and resulting peptides were analyzed by HPLC.

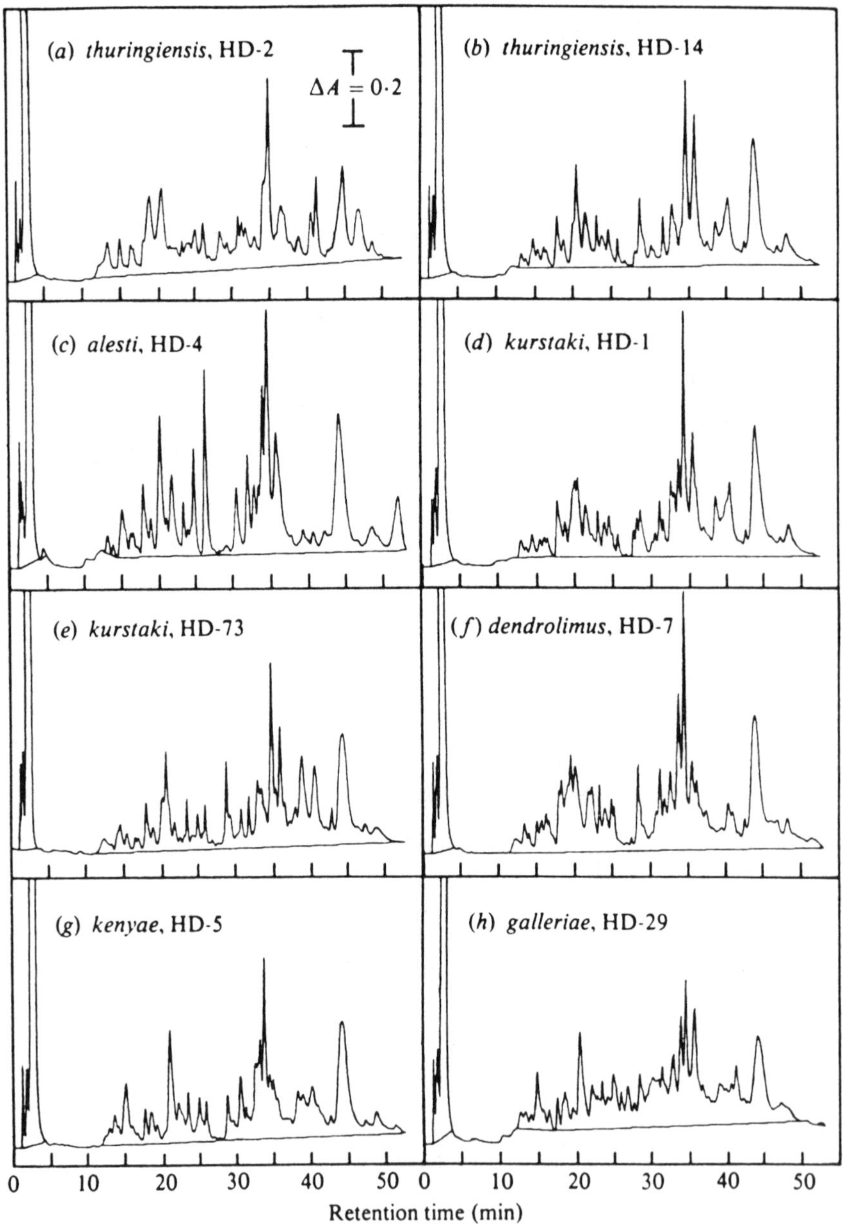

Figure 4. Peptide mapping of P1 isolated from various stains of B.t. P1 was isolated from the crystal by Sephacryl column chromatography. (Reproduced from ref. 24 by kind permission from the Society for General Microbiology.)

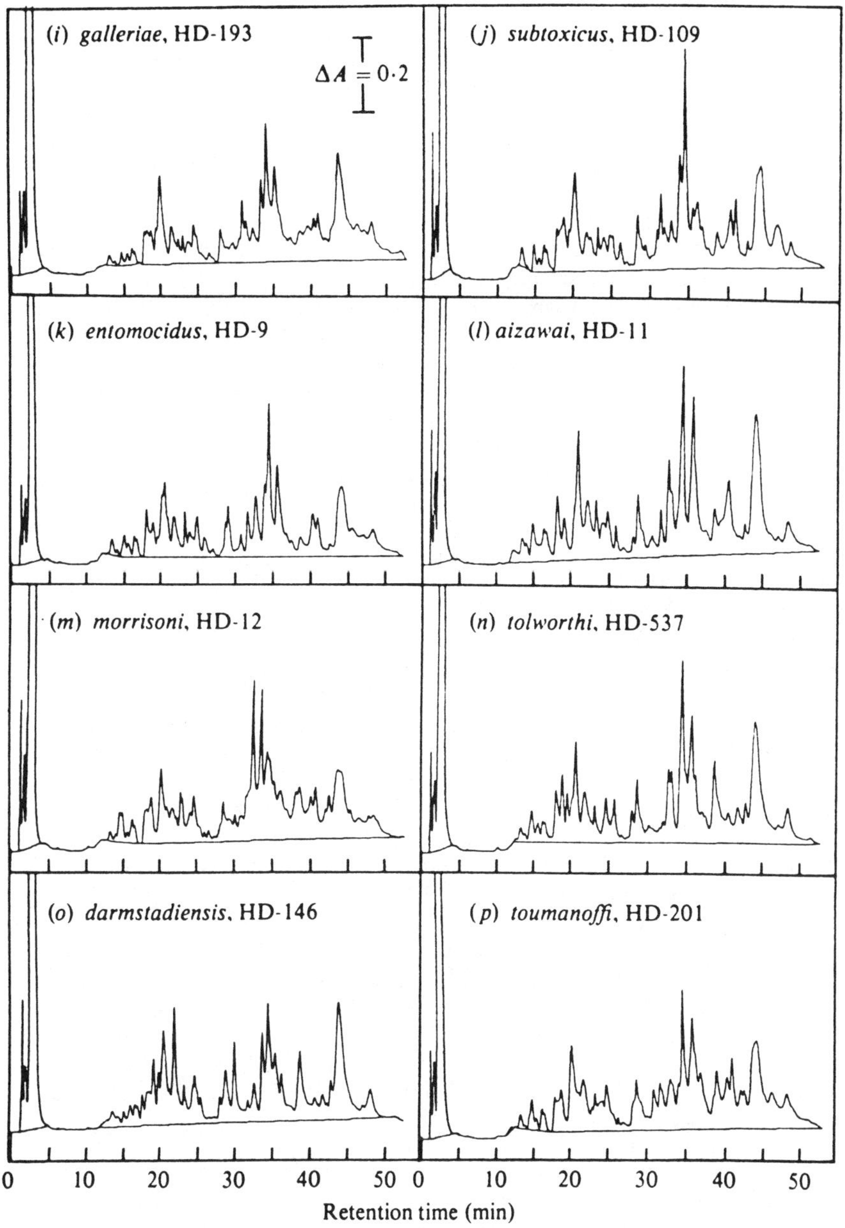

Figure 4. Continued.

Expression Levels of cryIA-type Genes

A large number of B.t. strains belong to subsp. kurstaki. According to USDA bioassay data (H. T. Dulmage, unpublished), they produce a variety of crystals having different insecticidal activity spectra. Using rocket immunoelectrophoresis, Yamamoto et al. (9) have shown a correlation between serological properties of crystals and their activity spectra. These observations were explained by a molecular biology finding that there are three genes termed cryIA in HD-1 and some other strains of B.t. kurstaki (25). Crystal proteins encoded by these genes are highly homologous but some differences in the amino acid sequence cause significant differences in the activity spectrum (18). Some, or perhaps many, kurstaki strains contain two or three cryIA genes (15). HPLC peptide mapping revealed that crystals of some strains were made of a mixture of proteins encoded by cryIA genes and the ratio of CryIA proteins varied from strain to strain (20). For example, a phenotype of a strain containing cryIA(a) and cryIA(b) genes is a crystal composed of individual phenotypes of these genes, CryIA(a) and CryIA(b) proteins. Differences in the ratio cause the differences in the activity spectrum.

In order to obtain individual finger prints of proteins encoded by cryIA genes, a number of plasmid-cured strains of B.t. kurstaki HD-1 and HD-263 were produced, and peptide maps of the P1 isolated from these strains were compared (20). It was fortunate that cryIA genes are on different plasmids, one or more of which could be cured. Figure 5 shows HPLC peptide maps of P1s from HD-1 and its plasmid cured derivatives. HD1-2 had lost the 44 megadalton (MDa) (67-kb) plasmid but retained the 110-MDa (167-kb) plasmid. On the other hand, HD1-9 had lost the 110-MDa plasmid but retained the 44-MDa plasmid. It was shown by mass spectroscopy that HD1-2 produces P1 encoded by cryIA(a) (20). As shown in Figure 5, there were noticeable differences in the map between HD1-2 and HD1-9. Furthermore, peptide maps of these HD-1 derivatives were also different from the map of the wild-type HD-1. It was assumed that cryIA genes on the 44-MDa and 110-MDa plasmids are expressed independently, and as a result, HD-1 produces a bipyramidal crystal made of mixed P1s. In order to determine expression levels of cryIA genes in the wild-type HD-1, P1s f rom HD1-2 and HD1-9 were purified separately, mixed in different ratios at 10% increments, trypsin-digested, and mapped by HPLC. As shown in Figure 5, a 50:50 mixture of P1s from HD1-2 and HD1-9 produced a pattern nearly identical to that of P1 from HD1-7. HD1-7 retains both 44 and 110-MDa plasmids and its peptide map was identical to that of the wild-type HD-1. The result indicates that half of the P1 crystal of HD-1 is composed of the CryIA(a) protein.

B.t. thuringiensis HD-14, kurstaki HD-1, tolworthi HD-537, and kenyae HD-5 were selected to compare P2 by HPLC peptide mapping (24). These strains were found to produce P2 when analyzed by immunoelectrophoresis. P2 was isolated by Sephacryl column chromatography. HPLC peptide mapping indicated that P2s from these four strains are quite different. Since two partly heterogeneous genes, cryIIA and cryIIB encoding P2 have been cloned, it is conceivable that several proteins encoded by different cryII genes compose the P2 crystal.

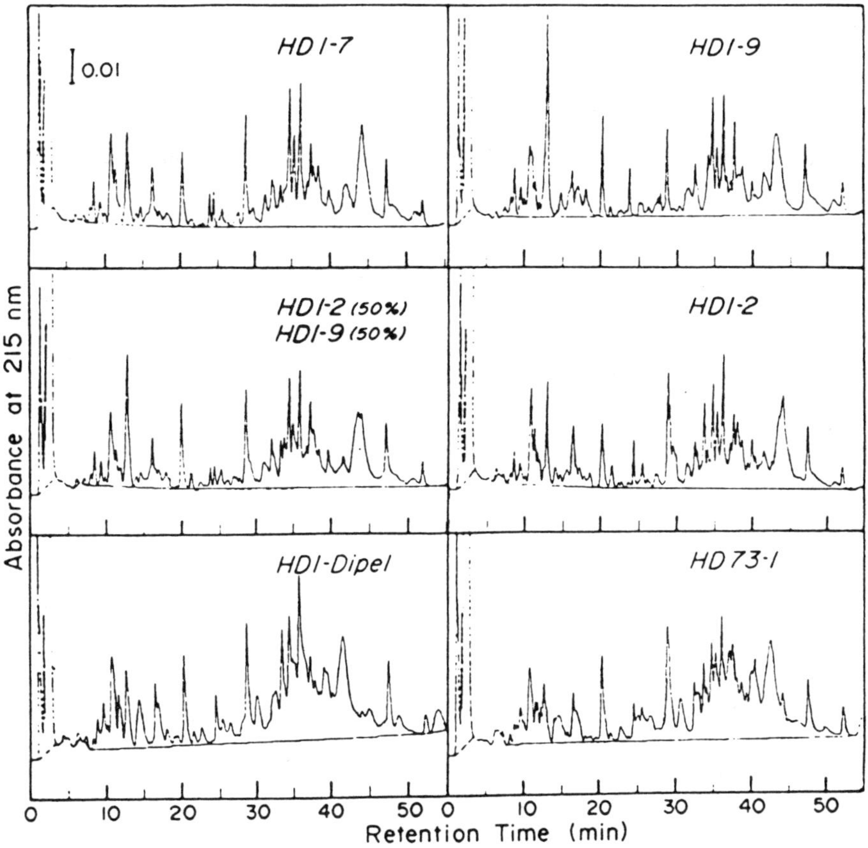

Figure 5. Peptide mapping of P1 isolated from B.t. kurstaki HD-1 and its derivatives. P1 was isolated from the crystal by Sephacryl column chromatography. (Reproduced from ref. 20 by kind permission from Springer-Verlag.)

Isolation of P1 by Affinity Chromatography

It was shown above that P1 can be purified by Sephacryl column chromatography. The purification, however, takes several hours to complete one sample. It is not practical to prepare P1s from many different B.t. strains with Sephacryl chromatography. In order to circumvent the problem, affinity chromatography utilizing an antibody immobilized on Sepharose was evaluated. Cloning and sequencing of cryIA-type genes revealed that they are highly homologous, in particular in their C-terminal domains (18). It is assumed that antibody made against the product (P1) of a cryIA-type gene cross-reacts with all other CryIA P1s.

Antiserum against P1 that had been purified from B.t. kurstaki HD-73 by Sephacryl column chromatography was produced. HD-73 is known to produce the CryIA(c) protein (13). It is also known that HD-73 does not produce P2 (9). To purify the antibody, the antiserum was first precipitated with 50%-saturated ammonium sulfate. The antibody, precipitated with ammonium sulfate, was dissolved in 10 mM Tris·HCl, pH 8, and further purified by antigen-immobilized affinity chromatography. P1 isolated from HD-73 by Sephacryl chromatography was coupled to Sepharose 4B by the CNBr activation method. The ammonium-sulfate fractionated antibody was applied onto the antigen immobilized column, and all unbound materials were eluted with 10 mM Tris·HCl, pH 8. After the gel was washed with 0.5 M NaCl, the antibody that bound to the antigen, P1 from HD-73, was eluted with 0.02 N NaOH. The eluate containing antibody specific to P1 of HD-73 was neutralized with HCl and 0.1 M $NaHCO_3$ and immediately immobilized on CNBr-activated Sepharose 4B.

Using this antibody-immobilized column, P1 was isolated from crystal extracts in about 1 hr. Figure 6 shows a typical chromatogram. In the experiment of Figure 6, HD-1 crystals were solubilized in 2% mercaptoethanol and NaOH at pH 10.5. Unlike the Sephacryl chromatography, the pH of the solubilized crystal must be neutralized using 25 mM Tris·HCl, pH 8, and dilute HCl. The antibody does not bind to the antigen under the conditions (pH 10.5) used to dissolve the crystals. Alternatively, P1 is selectively precipitated at pH 4.4. The acid-precipitated P1 can be collected by centrifugation and dissolved in 10 mM Tris·HCl, pH 8, before loading onto the affinity column. Conditions of the affinity chromatography for P1 (antigen) were the same as those for the antibody. P1 was eluted from the column with 0.02 N NaOH and neutralized with 100 mM Tris·HCl, pH 8. The affinity-purified P1 was suitable for HPLC peptide mapping. Its map was identical to that of Sephacryl-purified P1.

This affinity chromatography technique was tested with an E. coli clone expressing a B.t. cryIA gene, allowing a rapid isolation of P1. The purity of P1 from E. coli is equivalent to P1 isolated from the B.t. crystal. The antiserum, however, had to be pre-absorbed with non-recombinant E. coli cell lysate to remove antibodies that react with E. coli proteins.

Active Core of P1

P1 is a protoxin that is activated in the digestive juice of insects to yield the active core or the toxin (25). HPLC further demonstrated its utility in the

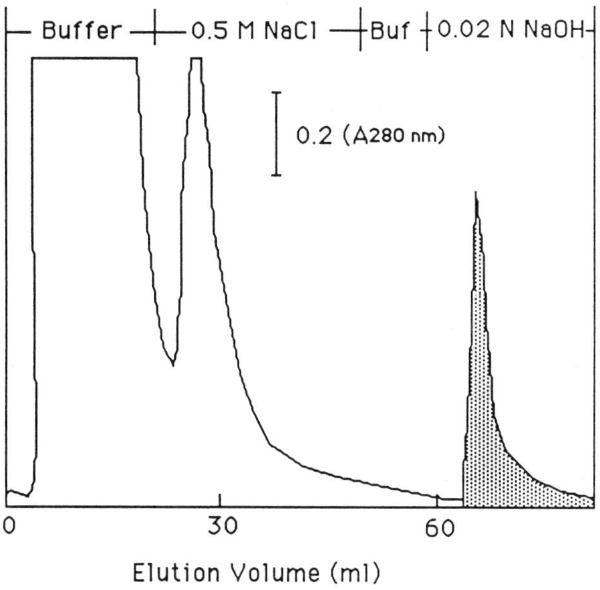

Figure 6. Isolation of P1 from the B.t. crystal by affinity chromatography. P1 was isolated from solubilized crystals by affinity chromatography using an antibody immobilized column. The antibody was made against P1 of HD-73 and purified by antigen-immobilized column chromatography. Buffer and Buf indicate 10 mM Tris·HCl, pH 8. The stippled peak contains the pure P1.

biochemistry of B.t. in the analysis of the active core of P1 isolated from HD1-2 (20). The active core is a proteinase-resistant portion of P1 (3). Nagamatsu et al. (4) isolated the active core from B.t. dendrolimus and determined its N-terminal amino acid sequence. The sequence of the active core corresponds to the sequence of P1 encoded by the cryIA gene starting at the 29th amino acid. The C-terminus of the core, however, has not been clearly determined. Schnepf and Whiteley (26) reported that the activity was lost when the cryIA(a) gene was truncated at 603 codons.

The following experiment (20) was performed for two purposes. One was to determine the type of P1 produced by HD1-2. It is known that HD-1 contains three partly homologous cryIA genes whose sequences have been determined (13, 17, 27). The sequencing revealed that most of the heterogeneities occur in the N-terminal half. In order to identify the cryIA gene that is expressed in HD1-2 as P1, P1 from HD1-2 was activated by trypsin, and the active core was analyzed by HPLC peptide mapping. The other purpose was to delineate the active core. To determine the C-terminus of the active core, it was necessary to locate a particular fragment around the 600 amino acid residue in a mixture of trypsin-digested core. P1 was purified from HD1-2, activated with trypsin in vitro, and the active core was isolated by ion-exchange chromatography using a MonoQ HR5/5 column. The active core was denatured in 8 M urea and digested with trypsin for the second time. Unlike the activation of P1, the second trypsin digestion was carried to completion. Peptides produced by the second trypsin digestion were separated by HPLC, and their molecular weights were determined by liquid matrix secondary ion mass spectrometry. Table II lists molecular weights of peptides derived from the second trypsin digestion of the active core. These molecular weights were compared with those calculated from factitious trypsin digestions of three different P1s encoded by cryIA-type genes. When a particular molecular weight determined by mass spectroscopy was found in a pool of calculated molecular weights based on the factitious digestion, numbers of amino acid residues corresponding to the start and end of a factitious tryptic peptide are listed in the table. The table clearly indicates that P1 of HD1-2 is a product of the cryIA(a) gene.

Molecular weight of the active core was estimated by SDS-PAGE as 64-kDa plus or minus 2-kDa. N-terminal amino acids of the core were sequenced and found as Ile-Glu-Thr-Gly-Tyr-Thr-Pro-- which is known to be the sequence starting at the 29th amino acid residue of P1 as deduced from the cryIA(a) gene sequence. In order to excise out the core having a molecular weight in the range of 62- to 66-kDa, only a few trypsin sites must be considered for the C-terminal end. Two of these sites are 600 Arg and 618 Arg for 64- and 66-kDa, respectively. As shown in Table II, the trypsin-digested core had a fragment of Mr=2157 that corresponds to the peptide from 601 to 618 amino acid residues of the CryIA(a) protein. Therefore, it was concluded that the active core of CryIA(a) is 66-kDa and extends from 29 Ile to 618 Arg.

Conclusion

Commercially important strains, such as HD-1 and B.t. israelensis, often carry several cry genes and some of them, if not all, are highly expressed. It is

Table II. Molecular weights of tryptic peptides derived from the proteinase-resistant core of the HD1-2 protein

	AA residue No.*		
Mr**	cryIA(a)	cryIA(b)	cryIA(b)
548	255-258	-	255-258
764	88-93	88-93	88-93
772	287-292	-	-
781	193-198	193-198	193-198
804	259-265	259-265	259-265
816	218-224	218-224	218-224
907	174-181	174-181	174-181
936	360-367	-	-
1038	210-217	210-217	210-217
1059	502-511	-	-
1089	490-500	-	-
1098	512-521	514-523	-
1129	449-457	-	-
1149	478-489	-	-
1234	182-192	182-192	182-192
1284	199-209	-	199-209
1398	116-127	116-127	116-127
1901	266-281	266-281	266-281
2157	601-618	603-620	-
2201	458-477	-	-

* Numbers of amino acid residues showing the start and end of each tryptic peptide.

** Molecular weights of tryptic peptides isolated from the core of the HD1-2 protein by HPLC.

\- Indicates that a peptide having the molecular weight determined by mass spectroscopy was not found in the pool of factitious tryptic peptides.

(Reproduced from ref. 20 by kind permission from Springer-Verlag.)

important to know which gene or genes are expressed and at what levels. It is also important to understand the crystal component of a newly isolated B.t. In this ACS symposium, I have reviewed applications of HPLC to characterize the B.t. crystal protein. HPLC offers several ways to investigate the biochemical and molecular biological nature of the crystal protein. It has been clearly indicated that HPLC peptide mapping is a powerful technique when an unknown crystal protein is compared with standards.

Literature Cited

1. de Barjac, H.; Bonnefoi, A. Entomophaga 1973, 18, 5.
2. Huber, H. E.; Lüthy, P.; Rudolf, H. -R.; Cordier, J.-L. Arch. Microbiol. 1981, 129, 14.
3. Lilley, M.; Ruffell, R. N.; Somerville, H. J. J. Gen. Microbiol. 1980, 118, 1.
4. Nagamatsu, Y.; Itai, Y.; Hakata, C.; Funatsu, G.; Hayashi, K. Agric. Biol. Chem. 1984, 48, 611.
5. Dulmage, H. T. J. Invertebr. Path. 1970, 15, 232.
6. Yamamoto, T; Iizuka, T. Arch. Biochem. Biophys. 1983, 227, 233.
7. Hall, I. M.; Arakawa, K. Y.; Dulmage, H. T.; Correa, J. A. Mosq. News 1977, 37, 246.
8. Yamamoto, T.; McLaughlin, R. E. Biochem. Biophys. Res. Commun. 1981, 103, 414.
9. Yamamoto, T.; Garcia, J. A.; Dulmage, H. T. J. Invertebr. Pathol. 1983, 41, 122.
10. Gonzalez, J. M. Jr.; Dulmage, H. T.; Carlton, B. C. Plasmid 1981, 5, 351.
11. Kronstad, J.W.; Schnepf, H.E.; Whiteley, H.R. J. Bacteriol. 1983, 154, 419.
12. Schnepf, H. E.; Whiteley, H. R. Proc. Natl. Acad. Sci. USA 1981, 78, 2893.
13. Adang, M. J.; Staver, M. J.; Rocheleau, T. A.; Leighton, J.; Baker, R. F. Gene 1985, 36, 289.
14. Held, G. A.; Bulla, L. A. Jr.; Ferrari, E.; Hoch, J.; Aronson, A. I.; Minnich, S. A. Proc. Natl. Acad. Sci. USA 1982, 79, 6065.
15. Kronstad, J. W.; Whiteley, H. R. J. Bacteriol. 1984, 160, 95.
16. McLinden, J. H.; Sabourin, J. R.; Clark, B. D.; Genslar, D. R.; Workman, W. E.; Dean, D. H. Appl. Environ. Microbiol. 1985, 50, 623.
17. Thorne, L.; Garduno, F.; Thompson, T.; Decker, D.; Zounes, M.; Wild, M; Walfield, A. M.; Pollock, T. J. J. Bacteriol. 1986, 166, 801.
18. Höfte, H.; Whiteley, H. R. Microbiol. Rev. 1989, 53, 242.
19. Angus, T. A. Can. J. Microbiol. 1956, 2, 416.
20. Yamamoto, T.; Ehmann, A.; Gonzalez, Jr., J. M.; Carlton, B. C. Curr. Microbiol. 1988, 17, 5.
21. Lecadet, M. M. C. R. Hebd. Seanc. Acad. Sci. 1966, 262, 195.
22. Fullmer, C. S.; Wasserman, R. H. J. Biol. Chem. 1977, 254, 7208.
23. Meek, J. L. Proc. Natl. Acad. Sci. USA 1980, 77, 1632.
24. Yamamoto, T. J. Gen. Microbiol. 1983, 129, 2595.
25. Kronstad, J. W.; Whiteley, H. R. Gene 1986, 43, 29.
26. Schnepf, H. E.; Whiteley. H. R. J. Biol. Chem. 1985, 260, 6273.
27. Schnepf, H. E.; Wong, H. C.; Whiteley. H. R. J. Biol. Chem. 1985, 260, 6264.

RECEIVED March 6, 1990

Chapter 7

Characterization of Parasporal Crystal Toxins of *Bacillus thuringiensis* Subspecies *kurstaki* Strains HD-1 and NRD-12

Use of Oligonucleotide Probes and Cyanogen Bromide Mapping

L. Masson, M. Bossé, G. Préfontaine, L. Péloquin, P. C. K. Lau, and R. Brousseau

Biotechnology Institute, 6100 Royalmount Avenue, Montreal, Quebec H4P 2R2, Canada

It has been shown that many strains of *Bacillus thuringiensis* produce more than one insecticidal crystal protein (icp). This includes the important commercial strain HD-1 and the newer NRD-12 strain, reported to be more effective than HD-1 against the forestry pest *Lymantria dispar* (gypsy moth). Oligonucleotide probes afford a rapid way to screen isolates of *Bacillus thuringiensis* for the presence or absence of the distinct icp gene types; they can also highlight subtle differences within icp gene types which escape detection by other means. Using these probes we found essentially identical genes between HD-1 and NRD-12. In order to quantify the various types of icps present within the parasporal crystals produced by these two commercial strains, a method employing cyanogen bromide (CNBr) has been developed. This method shows a small difference in gene expression levels between these two strains. It also establishes that, contrary to previous work, the *cryIA(c)* type gene is expressed to a significant degree in the HD-1 strain.

Although the current share of biological insecticides in the total insecticide market is generally evaluated at < 1% (1), this share is expected to increase substantially over the next several years. Biological insecticides provide better safety margins and lower exposure to the human population than the currently dominant chemical insecticides. The best known biological insecticide, *Bacillus thuringiensis*, has been steadily gaining superiority over its chemical competitors, particularly in the forestry market; this is largely due to its excellent environmental record. In the eastern part of North America the forestry programs have primarily been aimed at controlling the forest defoliators such as the gypsy moth *Lymantria dispar* and the spruce budworm *Choristoneura fumiferana*. These pests are of great importance to Canadian forestry, as reflected in the Canadian share of the overall *Bacillus thuringiensis* market, which may account for as much as one third to one half of the total worldwide market ($US 20 million out of a total of $US 60 million per year) depending on the severity of infestations in the individual provinces.

There is thus substantial economic importance in identifying strains of *Bacillus thuringiensis* which offer increased activity against forest pests. Such a strain was identified in 1984 as NRD-12 and has been studied extensively by several research groups (2). Recently a report appeared in which the NRD-12 strain failed to exhibit

0097–6156/90/0432–0061$06.00/0

higher activity in a field test (3). It is therefore important to compare the icp genes present in HD-1 and NRD-12 and also to be able to quantify the level of each of the various types of icps present in these strains.

The major part of the insecticidal activity of *Bacillus thuringiensis* resides in the insecticidal crystal proteins (icps) produced during sporulation. Several types of icps are known; they have been reviewed recently and assigned to groups such as *cryI*, *cryII*, etc (4). The icps which are exclusively toxic to lepidopterans are grouped in the *cryI* class; they consist of protoxins 130-133 kDa in size which combine to form parasporal bipyramidal crystals. Upon ingestion by a target insect, these protoxins are proteolytically cleaved in the digestive tract to form active toxins of 55-60 kDa. The major lepidopteran toxins found in the parasporal crystals of both HD-1 and NRD-12 belong to the *cryIA* group. The three subtypes are named *cryIA(a)*, *cryIA(b)* and *cryIA(c)*; they share substantial sequence homology both at the DNA and protein level.

The presence of many structurally related genes encoding different icps within the same isolate of *Bacillus thuringiensis* raises several important questions:

1. How many icp genes are present in a given strain of *Bacillus thuringiensis*?
2. Given that the icp genes are plasmid-borne (5), how stable are the strains in terms of the icp genes present?
3. In a multigene strain, what is the expression level of the individual genes?
4. Is there a synergistic or antagonistic effect in toxicity caused by the presence of different icps in a multigene strain?

Through the use of gene cloning, oligonucleotide probes and cyanogen bromide cleavage, we have attempted to answer some of these questions in the strains HD-1 and NRD-12.

Experimental Section.

Gene analysis. Cloning of the *cryIA* genes of HD-1 and NRD-12 was accomplished by standard recombinant DNA techniques (6). The in situ agarose gel hybridizations were performed according to a published procedure (7). The oligonucleotide probes were prepared on a commercially available automated DNA synthesizer (Applied Biosystems Model 380A) using cyanoethylphosphoramidite chemistry. The probes were labeled using ^{32}P γ-ATP and T4 polynucleotide kinase according to (6). The specific oligonucleotide probe used in Figure 1 to distinguish the *cryIA(c)* genes of HD-1 and NRD-12 from the HD-73 *cryIA(c)* gene had the sequence 5'-TTATCCTCTACTTTTTAT. Hybridization conditions for this probe were as follows: hybridization overnight at 37°C in 6x SSC/1x Denhardt's mix/5% dextran sulfate/0.05% Triton X100, followed by two 20 minute washes at 40°C in 2x SSC; autoradiography time was five days.

The general oligonucleotide probe used in Figure 2 to hybridize to all *cryIA* genes had the sequence 5'-CTAGTTGATATAATATGC. Hybridization conditions for this probe were as follows: hybridization overnight at 38°C in 6x SSC/1x Denhardt's mix/1% dextran sulfate/0.05% Triton X100, followed by four 15 minute washes at 40°C in 3x SSC; autoradiography time was 24 hours.

Cyanogen Bromide Analysis. The HD-1 and NRD-12 crystals were a gift from Tim Lessard, NRC, Ottawa and had been purified as described (8). Inclusion bodies from recombinant *Escherichia coli* were harvested by centrifugation of insoluble material after lysis in a French pressure cell (two passes at 11,000 psi internal pressure). The inclusion bodies, resuspended in water, were then pelleted through a renograffin gradient (1 part cell suspension:1.8 part renograffin, 30,000xg, 1 hour in a Beckman SW28 rotor). After rinsing the inclusion body pellet with water and repelleting at 10000xg for 15 minutes, the inclusion bodies were recovered and were consistently found to contain 90-95% protoxin as judged by SDS-PAGE.

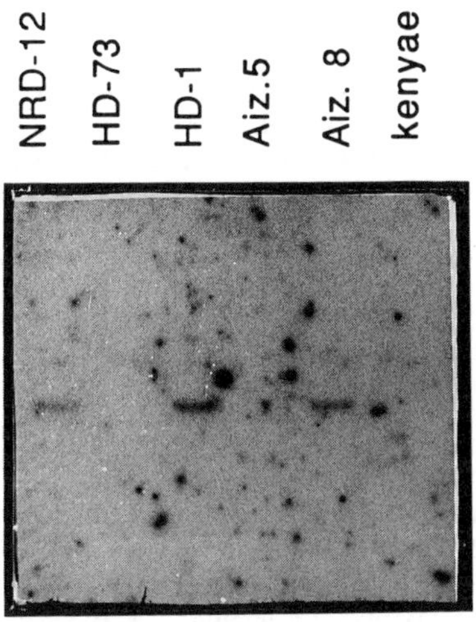

Figure 1. In situ hybridization of an agarose gel containing *HindIII* fragments of DNA isolated from six different *Bacillus thuringiensis* strains with an oligonucleotide probe specific for the *cryIA(c)* gene variant found in the NRD-12 strain.

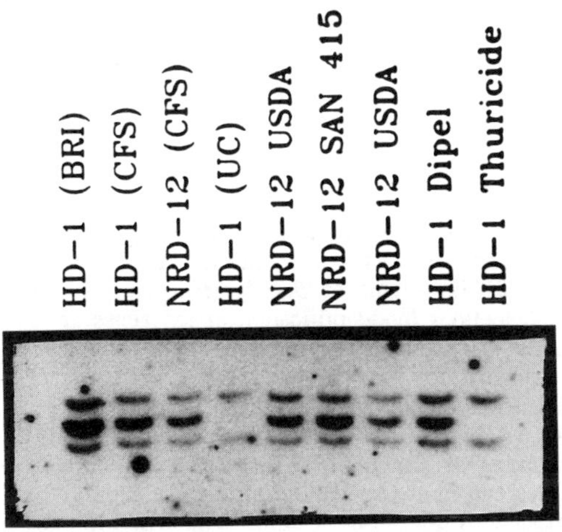

Figure 2. In situ hybridization of an agarose gel containing *HindIII* fragments of DNA isolated from several cultures of HD-1 and NRD-12 found in various laboratories. The probe was an oligonucleotide chosen to hybridize all *cryIA* genes. BRI = Biotechnology Research Institute (Montreal); CFS= Canadian Forestry Service, Sault Sainte Marie, Ontario; UC= University of California at Riverside; USDA=NorthEastern Forest Experiment Station, Hamden, CT; SAN 415, Dipel and Thuricide refer to commercial formulations.

Protoxins from either inclusion bodies or parasporal crystals were dissolved in 50 mM sodium carbonate buffer pH 9.5 containing 10 mM 2-mercaptoethanol at 37°C for one hour. After centrifugation of insoluble proteins (including any *cryII* toxins in the parasporal crystals), the solubilized proteins were concentrated and lyophilized to remove the 2-mercaptoethanol. The lyophilized proteins were redissolved in 70% (v/v) aqueous formic acid and a large excess of cyanogen bromide (500-fold over the calculated amount of methionine residues present in the protoxins) was added. After 24-30 hours at 37°C, the cleavage products were diluted with water and lyophilized three times to remove the cyanogen bromide and formic acid. The peptides were then resuspended in SDS sample buffer, boiled for five minutes and loaded on a 10% (w/v) SDS-polyacrylamide gel for electrophoresis. Quantitation of the "signature" peptides was accomplished by scanning the dried polyacrylamide gels (stained with Coomassie Blue) using an LKB Ultro-scan XL laser densitometer.

CF-1 insect cell assay. This assay relies on the ability of activated protoxins to kill cultured CF-1 cells (L. Gringorten, manuscript submitted). CF-1 cells (from *Choristoneura fumiferana* neonate larvae trypsinized cells, obtained from Dr. S. Sohi, Canadian Forestry Service, Sault Sainte Marie, Ont.) were maintained on Grace's insect cell medium supplemented with 10% Fetal Bovine Serum and 0.25% Tryptose. A lawn of insect cells was prepared by mixing 2 x 10^8 cells in four mls of saline buffer with four mls of 2% Seaplaque agarose at 37°C (FMC Corporation, Rockland, USA) and pouring the suspension over a 1% agarose base into the cover of a 96-well microtiter plate. The protoxins were activated by treating either the solubilized parasporal crystals or *Escherichia coli* inclusion bodies with 1% (v/v) *Bombyx mori* gut juice (in 0.2M CAPS buffer pH 10.5 + 0.02% DTT) at 28°C overnight. Two-fold dilutions of the activated toxins were spotted onto the surface of the insect cell lawn and allowed to react for one hour at room temperature. The lawn was then stained with 0.2% trypan blue and destained in buffered saline, then in 1.34% KCl. Toxicity thresholds were determined as the lowest concentration of activated protoxin to produce a visible lytic spot on the lawn surface.

Results and Discussion

Use of Oligonucleotide Probes to Study icp Genes Contained in Commercial Strains HD-1 and NRD-12. Oligonucleotides have already been used to study the occurrence of icp genes among fifteen strains of *Bacillus thuringiensis* (9). Briefly stated, the advantages of oligonucleotides for identification of icp genes are as follows:

1. Oligonucleotides are now readily available from commercial DNA synthesizers and can be easily prepared for any published DNA sequence. As an example, using colony hybridization with oligonucleotides designed from the published sequences of *cryIA(a)* (10) *cryIA(b)* (11) and *cryIA(c)* (12), we were able to distinguish and identify all three types of clones from a single shotgun cloning experiment.
2. Oligonucleotides can be made very specific for a small, characteristic region of a large DNA sequence, thus revealing features which would not be apparent with a larger DNA probe. Figure 1 exemplifies this specificity. After it was discovered by one of us (M. Bossé) that the DNA sequence of the *cryIA(c)* of the HD-73 strain was slightly different from the *cryIA(c)* genes of the HD-1 and NRD-12 genes, an oligonucleotide was designed and synthesized which would specifically hybridize with the latter two but not the former. The same experiment also established the presence of the HD-1 type of *cryIA(c)* in one of two *aizawai* strains and its absence in a *kenyae* strain. With other DNA probes, both *aizawai* strains gave a three band *HindIII* pattern typical of the *cryIA(a)*, *cryIA(b)* and *cryIA(c)* genes found in HD-1; yet they can be distinguished by this new oligonucleotide.
3. The "one band, one gene" advantage. On a Southern blot, the number of bands hybridizing with an oligonucleotide is directly related to the number of genes

homologous to that particular DNA sequence. If a particular gene is cut in two by a restriction enzyme located in the probe area, a long DNA probe will give two bands for one gene. In contrast, an oligonucleotide will be unable to hybridize near the stringent temperature unless essentially all of its bases hybridize to the same DNA piece. In this case, two bands can only arise if two distinct genes are present.

The presence of several distinct icp genes in the standard commercial strain HD-1 has been recognized for several years (13). It has also been known for several years that the *cryIA(b)* of HD-1 could be lost under certain conditions (1). This loss presumably accounts for the difference in activity ratios between the HD-1-1971-S standard and the HD-1-1980-S standard for Biological International Unit (BIU) definition (1). This situation emphasizes the need for applying the newer analytical methods of genetic analysis to a prospective standard, so as to gain additional insurance against the inherent variability of living organisms.

As part of our study of NRD-12, we discovered early that it, like HD-1, is a three gene strain containing the *cryIA(a)*, *cryIA(b)* and *cryIA(c)* gene types (9). Extensive studies using restriction enzyme mapping, oligonucleotide hybridization and DNA sequencing have established that the three *cryIA* genes present in NRD-12 are essentially indistinguishable from those of HD-1 (14) . It has also been found that the unavoidable variability of insect bioassays between different laboratories makes it very difficult to pin down the genetic features of the NRD-12 strain responsible for the higher level of activity reported. Some groups have failed to detect any increased activity of NRD-12 over HD-1 in field trials (3) or in the laboratory (J. Cabana, personal communication). The possibility exists that HD-1 and NRD-12 are essentially identical strains with respect to their *cryIA* genes, and that they may differ only in other aspects which have a lesser impact on toxicity such as spore virulence, β-exotoxin levels, phospholipase or chitinase levels.

Being aware of the possibility of the HD-1 strain losing its *cryIA(b)* gene, we have analyzed the HD-1 and NRD-12 strains from our various collaborators, using oligonucleotide hybridization (Figure 2). These results clearly establish that two of the HD-1 strains used in the comparison studies were two gene strains whereas others were three gene strains. The loss of the *cryIA(b)* gene may explain the variability of the toxicity results, inasmuch as researchers using a three gene HD-1 reference find small difference or no difference at all between HD-1 and NRD-12.

In an attempt to resolve the issue, we have cloned and expressed in *Escherichia coli* all six *cryIA* genes from HD-1 and NRD-12 (Masson et al., manuscript submitted) so that bioassays could be performed on each individual gene product. Insect bioassays of these materials are currently under way.

Use of Cyanogen Bromide to Quantify the Levels of Individual Gene Products in the HD-1 and NRD-12 strains. DNA probes give information about the presence or absence of a particular gene class in a given *Bacillus thuringiensis* isolate; however, they give no information as to whether or not the protein encoded by this gene is actually expressed. It has been reported (13, 15) that certain *Bacillus thuringiensis* isolates contain icp genes carried on the bacterial chromosome, as opposed to plasmids, and that these chromosomal genes are not expressed. It has also been found recently (L. Masson, unpublished observation; M. Geiser, personal communication) that rearranged, presumably inactivated icp genes are present in certain *Bacillus thuringiensis* isolates. It is therefore essential to find a way to quantify directly the various types of icp(s) which may be present in the parasporal crystals of a given isolate.

Here again our initial work has focused on the commercial isolates HD-1 and NRD-12. It is quite possible that NRD-12 may possess higher activity than HD-1 against certain insects, even if the icp genes it contains were virtually identical to those of HD-1. This situation could arise if the relative levels of expression of the three icp *cryIA* genes present in both strains were to differ significantly between the strains. If these expression levels can be influenced by the growth conditions and/or the growth media, this could also explain the difference in toxicity results between laboratories or between the laboratory and the fermentation plant.

Calabrese and Nickerson (16) have used SDS-PAGE to demonstrate the presence of both 130 kDa and 133 kDa protoxins in parasporal crystals of the HD-1 strain. More recently an HPLC study has been conducted on proteolytic fragments from the icps of HD-1 (17); the conclusion was that the *cryIA(c)* gene was either not expressed or expressed at a low level (10% or less) in this strain. To our knowledge no method of quantitation applicable to a three gene strain has yet been published.

As mentioned earlier, we have cloned and expressed in *Escherichia coli* all six icp genes contained in HD-1 and NRD-12. In addition, we have fully sequenced the *cryIA(b)* from NRD-12 and have done extensive hybridization, restriction enzyme mapping and sequencing on the other five genes present in these strains. The *cryIA(a)* and *cryIA(b)* genes of HD-1 have been fully sequenced by other workers (10, 11). From this basis, we have developed a method based on SDS-PAGE electrophoresis of cyanogen bromide fragments which allows us to readily quantify the level of expression of each type of icp both in HD-1 and NRD-12.

The method relies on quantitation, by laser densitometry, of unique cyanogen bromide cleavage fragments created from the mixture of the three icps present in the P1 parasporal crystals of these strains. As reference materials, we used the purified individual icps from both strains cloned and expressed in *Escherichia coli*; this insures that there are no unexpected interferences from the two NRD-12 icp genes (*cryIA(a)* and *cryIA(c)*) which have not been completely sequenced. The calculated molecular weights of the largest CNBr peptides, which we refer to as "signature" peptides, are shown in Table I.

Table I. Position and predicted molecular weights of the "signature" peptides for the three *cryIA* classes of icps present in HD-1 and NRD-12

Gene class	Signature peptide region (#aa-#aa)	Expected molecular weight (kDa)
cryIA(a)	555-927	42.6
cryIA(b)	556-902	39.6
cryIA(c)	453-927	53.7

A complete cleavage of all the Met sites present within the icps is not particularly necessary. We aim to achieve a degree of cleavage sufficient so that the signature peptide is the largest fragment visible on the SDS-PAGE gel, since the presence of incompletely cleaved, low molecular weight peptides has no bearing on the total protoxin as calculated from the signature peptides. The result of one such gel is shown in Figure 3, which also shows the result of cyanogen bromide cleavage on three individual icps (produced in *Escherichia coli* as inclusion bodies) used to calibrate the method. It is noteworthy that the observed molecular weights of the signature peptides are approximately 12% higher than the calculated values. The reasons for this apparent anomaly are currently unknown; however, the ratios calculated for the three classes of protoxins remain the same whether one uses the apparent or the calculated molecular weights. Reproducibility of the method falls within a 2% to 5% standard deviation for the total crystal composition.

The numerical results of the method are shown in Figure 4 for the HD-1 versus NRD-12 comparison. It can be readily seen that, contrary to earlier results (17), the *cryIA(c)* gene product is abundantly present in both the HD-1 and NRD-12 parasporal crystals, representing approximately one-third of the total in both cases. The HD-1 strain seems to produce more *cryIA(b)* protein, whereas the NRD-12 has more

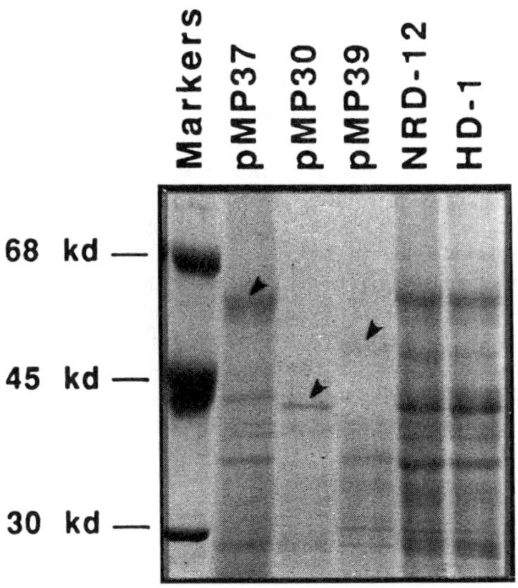

Figure 3. SDS-PAGE separation of the cyanogen bromide cleavage products from HD-1 parasporal crystals, NRD-12 parasporal crystals and the inclusion bodies from *Escherichia coli* producing isolated *cryIA* protoxins. The pMP39, pMp30 and pMP37 plasmids produce the NRD-12 *cryIA(a)*, *cryIA(b)* and *cryIA(c)* protoxins respectively.

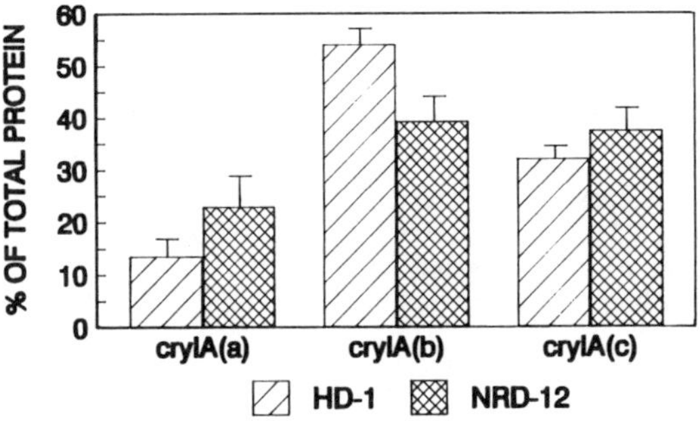

Figure 4. Composition of the HD-1 and NRD-12 parasporal crystals in terms of the three classes of *cryIA* proteins.

cryIA(a). These differences are reproducible for laboratory grown samples; whether they would also be found in commercial formulations remains to be seen. Investigation of the possible effects of growth medium on crystal content of the three gene types is also worthy of future investigation, inasmuch as such variation could affect efficacy of the commercial products from batch to batch.

In vitro Bioassays of the Individual Gene Products of HD-1 and NRD-12. Since our results are quite different from those previously published with regards to the expression of the *cryIA(c)* gene in HD-1 (17), we have sought to confirm our results through the use of an in vitro bioassay. The "lawn" assay (L. Gringorten et al., manuscript submitted) on the CF-1 cell line, established from neonate larvae of *Choristoneura fumiferana*, turned out to be very convenient for this purpose. This cell line is much more sensitive to the *cryIA(c)* gene product (approximately four orders of magnitude) than to the *cryIA(b)* and *cryIA(a)* gene products. We also established, using mixtures of icps produced individually in recombinant *Escherichia coli*, that there is neither synergy nor antagonism between the three types of gene products present in either HD-1 or NRD-12 with respect to toxicity towards CF-1 cells (data not shown). It is therefore possible to correlate directly the amount of *cryIA(c)* gene product present in the HD-1 and NRD-12 crystals with the threshold of toxicity observed with these crystals, compared to the threshold of toxicity of the individual gene products expressed in *Escherichia coli*. By this method we estimated (Table II) that approximately one-third of both the HD-1 and NRD-12 crystals are made up of the *cryIA(c)* gene product, an estimate which agrees well with the cyanogen bromide data discussed above.

Table II: Estimate of the *cryIA(c)* gene product levels in the P1 parasporal crystals of *Bacillus thuringiensis* strains HD-1 and NRD-12 using the CF-1 cell assay

Sample	Average threshold of toxicity to CF-1 cells (ng)	% Relative toxicity* (= % *cryIA(c)* in mixture)
HD-1 crystals	0.47	34
NRD-12 crystals	0.38	37
HD-1 *cryIA(c)* protoxin**	0.16	100
NRD-12 *cryIA(c)* protoxin**	0.14	100
NRD-12 *cryIA(b)* protoxin**	1000	0.02
NRD-12 *cryIA(a)* protoxin**	non-toxic	-

* The % relative values are calculated as the threshold toxicity of the *cryIA(c)* protoxin x 100 divided by the threshold toxicity of the crystal. All protoxin values represent the average of ≥ four separate experiments.

** Purified inclusion bodies from recombinant *Escherichia coli*.

Conclusion

We have established that NRD-12 and HD-1 are very similar in terms of the types of the three *cryIA* icp genes present. We do find a difference in the relative expression levels of the three gene types between HD-1 and NRD-12; further bioassay work is necessary to establish whether this difference leads to meaningful differences in activity against forestry insects. We have also demonstrated that, in contrast to previously published work, all three types of *cryIA* icps are expressed to a significant level in HD-1.

Legend of Symbols.

Abbreviations used: CAPS, 3-cyclohexylamino-1-propanesulfonic acid; CNBr, cyanogen bromide; DTT, dithiothreitol; SDS-PAGE, sodium dodecyl sulfate-polyacrylamide gel electrophoresis;

Literature Cited.

1. Wilcox, D.R.; Shivakumar, G.; Melin, B.E.; Miller, M.H.; Benso, T.A.; Schopp, C.W.; Casuto, D.; Gundling, G.J.; Bolling, T.J.; Spear, B.B.; Fox, J.L. In Protein Engineering: Applications in Science, Medicine and Industry; Academic Press, 1986; Chapter 25.
2. Dubois, N.R. In Proceedings on microbial control of spruce budworms and gypsy moth, United States Department of Agriculture, Broomall, PA. 1985; pp. 99-102.
3. Stelzer, M.J.; Beckwith, R.C. J. J. Econ. Entomol. 1988, 81, 880-886.
4. Höfte, H.; Whiteley, H.R. Microbiol. Rev. 1989, 53, 242-255.
5. Gonzalez, J.M. Jr.; Carlton, B.C. Plasmid 1980, 3, 92-98.
6. Maniatis, T.; Fritsch, E.F.; Sambrook, J. Molecular cloning: a laboratory manual Cold Spring Harbor Laboratory Press, 1982.
7. Meinkoth, J.; Wahl, G. Anal. Biochem. 1984, 138, 267-284.
8. Carey, P.R.; Fast, P.; Kaplan, H.; Pozsgay, M.; Biochim. Biophys. Acta 1986, 872, 169-176.
9. Préfontaine, G.; Fast, P.; Lau, P.C.K.; Hefford, M.A.; Hanna, Z.; Brousseau, R. Appl. Environ. Microbiol. 1987, 53, 2808-2814.
10. Schnepf, H.E.; Wong, H.C.; Whiteley, H.R. J. Biol. Chem. 1985. 260, 6264-6272.
11. Geiser, M.; Schweitzer, S.; Grimm, C. Gene 1986, 48, 109-118.
12. Adang, M.J.; Staver, M.J.; Rocheleau, T.A.; Leighton, J.; Barker, R.F.; Thompson, D.V. Gene 1985, 36, 289-300.
13. Kronstad, J.W.; Schnepf, H.E.; Whiteley, H.R. J. Bacteriol. 1983, 154, 419-428.
14. Hefford, M.A.; Brousseau, R.; Préfontaine, G.; Hanna, Z.; Condie, J.A.; Lau, P.C.K. J. Biotechnol. 1987, 6, 307-322.
15. Klier, A.; Parsot, C.; Rapoport, G. Nucleic Acids Res. 1983, 11, 3973-3987.
16. Calabrese, D.M.; Nickerson, K.W. Can. J. Microbiol. 1980, 26, 1006-1010.
17. Yamamoto, T.; Ehmann, A.; Gonzalez, J.M. Jr.; Carlton, B.C. Curr. Microbiol. 1988, 17, 5-12.

RECEIVED February 19, 1990

Chapter 8

Development of a High-Performance Liquid Chromatography Assay for *Bacillus thuringiensis* var. *san diego* δ-Endotoxin

Leonard Wittwer[1], Denise Colburn[2], Leslie A. Hickle[2], and T. G. Sambandan[2]

[1]Invitrogen Corporation, 1158 Sorrento Valley Road, Suite 20, San Diego, CA 92121
[2]Mycogen Corporation, 5451 Oberlin Drive, San Diego, CA 92121

A Reverse Phase HPLC assay has been developed for quantifying the delta-endotoxin produced by *Bacillus thuringiensis* var. *san diego*. The effects of ethyleneglycol as a mobile phase modifier and in stabilizing solubilized delta-endotoxin are discussed.

Bacillus thuringiensis var. *san diego* (BTSD) is a naturally occurring bacillus which was isolated at Mycogen Corporation. It produces a protein delta-endotoxin as a crystalline inclusion body which is active against certain coleopteran insects such as Colorado Potato Beetle and Elm Leaf Beetle (1.2). This protein is the active ingredient in M-One® insecticide. An analytical assay is needed to quantify the native toxin levels in both fermentation batches and final product formulations.

Currently the toxin is measured by SDS-PAGE (3) followed by laser densitometry. Further, to confirm the assay, a bioassay is also performed. This technique, although very sensitive, is not selective. SDS-PAGE measures total protein of a given molecular weight and does not differentiate between the active and the non-active forms of the endotoxin protein. As such, the assay has been difficult to correlate with the Bioassay (4). Furthermore, the assays were cumbersome, time-consuming, and the results can vary widely (±10% Standard Deviation). Hence, there was a need for a faster and more accurate assay to measure the endotoxin.

Reverse Phase HPLC has gained popularity as a tool for quantitation and purity analysis of proteins. The mechanism of separation on the micro hydrocarbon bonded silica particulates is essentially hydrophobic interaction. This technique can separate proteins of similar molecular weights and other hydrophobic proteins. The purpose of this investigation was to develop a stability-indicating high performance liquid chromatography (HPLC) method for the quantitation of endotoxin in both the broth and in the final product.

Materials and Methods

Preparation of Endotoxin. Standard toxin used was made by isolation and recrystalization from BTSD grown in production media. The fermentation broth was concentrated three-fold either by centrifugation or ultrafiltration. The pH of the concentrate was raised to 12 - 12.5 with 1 N NaOH. The mixture was stirred at room temperature for 20 minutes to dissolve the

0097–6156/90/0432–0070$06.00/0

toxin. Insoluble material was removed by centrifugation at 3500 RPM for 20 minutes. The supernatant was dialyzed against Tris- (50 mM Tris pH 7.5, 1 mM EDTA) for 2 hours at 4°C. As the pH of the sample dropped, the toxin precipitated inside the dialysis tube. The precipitated toxin was collected and washed with water until the resulting pellet was white and the supernatant was clear.

On SDS-PAGE the crystals yield a single band with an "apparent" molecular weight of about 64 kD. The crystalized toxin standard is stable on SDS, Page, as well as by bioassay for at least six months when stored in 10 mM Tris-HC1 buffer pH 7.5 at 4°C.

Sample Preparation for HPLC. The toxin standard was prepared by dissolving the toxin crystals in 10 mM NaOH and 1 mM Dithioerythritol (DTT). Ethylene glycol was added to 50% U/V and incubated at 100°C for 10 minutes. The protein content was determined by a Lowry assay using BSA as the standard. It should be noted that the determination of M-One protein by using BSA as standards would contribute to a systematic error. We chose to neglect it for its contribution was small compared to other experimental errors.

HPLC Equipment and Reagents. A Water's gradient system was used (680 gradient controller, Two - 510 pumps, U6K injector, 410 (UV) flow Detector). Data was collected with Gilson's 712 software package on a Dell System 310 computer. The column used was an Alltech Macrospere C4 (25CM 5 um particle size, 300 A pore). Detection was set at 254 nM and solvent flow was 1 ml/mm. Mobile phases A and B were water and acetonitrile (ACN) respectively. Both eluants were modified with ethylene glycol and Trifluoroacetic acid (TFA) as described in the text. All reagents were HPLC grade.

Results and Discussion

Preparation of endotoxin solution from the crystal suspension was difficult. The crystal did not dissolve easily in mild solvents. The endotoxin dissolved in mobile phase (40% water 60% acetonitrile 0.1% TFA) would form a gel after a few minutes at room temperature. Crystals were also difficult to dissolve in urea or guanidine. Further alkali solubilized toxin was unstable.

However, a solution of stable endotoxin was prepared by adding an equal volume of modified sample mix (2% SDS 10% Glycerol 2.5% Mercaptoethanol) to alkali solubilized endotoxin and heating the mixture at 100°C for 10 minutes. The resulting solution was stable and was suitable for HPLC analysis.

As with most protein analyses, a gradient HPLC Method was developed to analyze the M-One standard. Preliminary experiments with mobile phases also indicated that TFA was essential for the elution of endotoxin.

Although endotoxin gave the two characteristic peaks, ghost peaks were also encountered. For example, in the experiment when the standard was run with the following conditions isocratic 40% B for 10 minutes and a gradient to 70% B for 30 minutes and a return to 40% B in 5 minutes, a third peak appeared during the return to initial conditions. The peak was analyzed by SDS-PAGE and was found to be 64 kD endotoxin. To minimize ghosting and to enhance recovery, addition of ethylene glycol to the mobile phase was implemented. As shown in Figure 1, an addition of 2% ethylene glycol to the mobile phase eliminated the late peak and ghost peak in the blank runs.

The two peaks of endotoxin raised questions of purity of the product. Are two different proteins present or do the proteins exist as two forms (active and the non-active form)? To answer these questions, the various peaks were collected and were run on SD-PAGE. Both the peaks showed identical bank patterns to that of M-One, suggesting that the endotoxin existed in two forms.

To study the various forms, a thermal study was undertaken. The experiment was conducted on a sample that was made in 50% glycerol. As shown in Figure 2, the first peak was

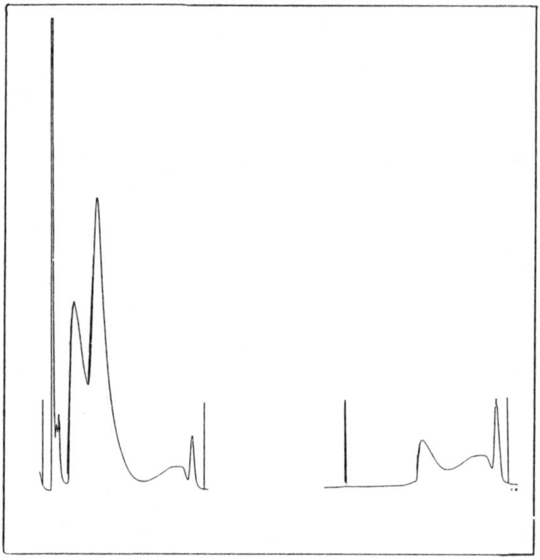

Figure 1A: Effect of Ethylene Glycol in the Mobile Phase

Mobile Phase:	A - Water, 0.015% TFA
	B - Acetonitrile, 0.015% TFA
Gradient:	10 minutes at 45% B, gradient to 90% B in 30 minutes
Sample:	BTSD crystals, dissolved in NaOH, SDS

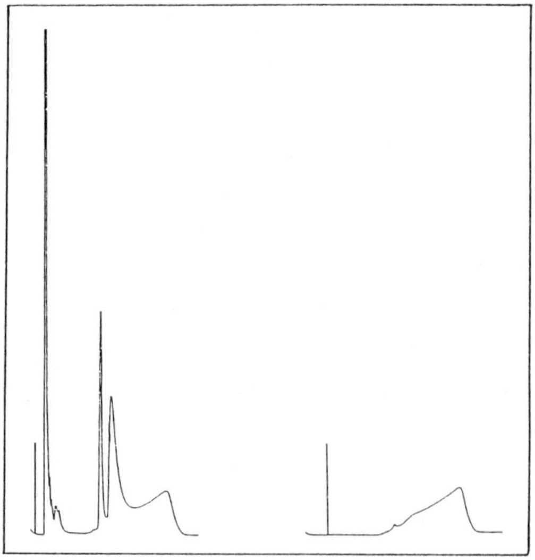

Figure 1B: Effect of Ethylene Glycol in the Mobile Phase

Mobile Phase:	A - Water, 2% ethylene glycol, 0.015% TFA
	B- Acetonitrile, 2% ethylene glycol, 0.015% TFA
Gradient:	10 minutes at 40% B, gradient to 70% B in 30 minutes
Sample:	BTSD crystals, dissolved in NaOH, SDS

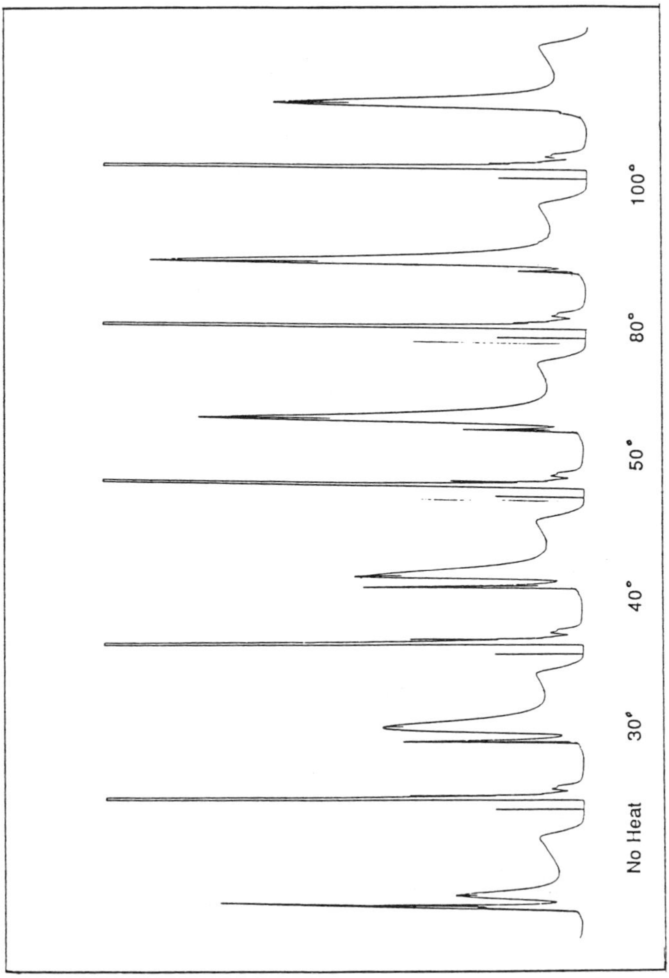

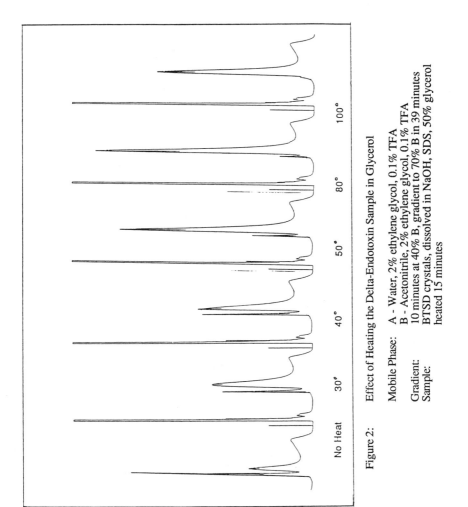

Figure 2: Effect of Heating the Delta-Endotoxin Sample in Glycerol

Mobile Phase: A - Water, 2% ethylene glycol, 0.1% TFA
B - Acetonitrile, 2% ethylene glycol, 0.1% TFA

Gradient: 10 minutes at 40% B, gradient to 70% B in 39 minutes

Sample: BTSD crystals, dissolved in NaOH, SDS, 50% glycerol heated 15 minutes essentially being converted to the second peak by just heating the sample. In fact, only one form, the second peak, was observed when the sample was heated to 100°C, indicating the second peak is a denatured form of endotoxin. Clearly, this assay could be used as a stability-indicating assay.

Preliminary studies with bioassay indicate that the native protein was indeed active and the others had minimal activity. The native protein could also be of different molecular weight.

During the course of this investigation, it was also discovered that the denatured protein could be renatured by addition of ethylene glycol. As shown in Figure 3, using the same experiment except substituting ethylene glycol for 50% glycerol, the sample returns to the native state on heating at 100°C. We have no explanation for this behavior. Further work is underway to study the mechanism of renaturation.

Various dilutions of the toxin standard were injected to produce a calibration curve based on peak area (Figure 4). Multiple 50 ul injections of the same protein concentration yielded peak areas with a coefficient of variation 0.05%.

Toxin Quantitation

Samples of formulated BTSD toxin (M-One, Mycogen Corporation) were assayed by comparison to the calibration curve. The formulation was diluted with water and centrifuged to recover the toxin crystals.

Chromatograms of this material yielded both peaks characteristic of the standard toxin injections. No combination of heat, ethylene glycol SDS, or glycerol converted the formulated toxin to one peak. Rather than assume the peak areas were additive, the quantitation scheme shown in Figure 5 was employed to measure toxin levels. A comparison of toxin values obtained from HPLC and SDS-PAGE assays are presented in Table I.

Table I. Comparison of HPLC and Gelscan Results on a Typical Formulated Sample

Sample	HPLC	Gelscan
A	5.0 X 10^3 ug/g	5.4 X 10^3 ug/g
B	6.2 X 10^3ug/g	6.1 X 10^3ug/g
C	5.0 X 10^3 ug/g	5.3 X 10^3ug/g

The difference between the values is within the expected error of the Gelscan assay.

More work to address, precision and recovery studies, and a more rigorous correlation of the HPLC assay to quantitation by SDS-PAGE is underway.

Summary

In this paper we have described the development of an assay for *Bacillus thuringiensis* var. *san diego* toxin using RP-HPLC. The protein is dissolved by raising the pH and then is stabilized by heating in the presence of 50% ethylene glycol. Calibration curves based on peak area of purified toxin are linear and reproducible. The assay requires only a few hours, replacing an SDS-PAGE based assay which ran overnight.

Two aspects of the assay are especially interesting. First is the increased recovery from the column by the addition of ethylene glycol to the mobile phase. The mechanism is unclear, but perhaps the glycol masks residual silanol groups or inhibits protein interaction with the bonded

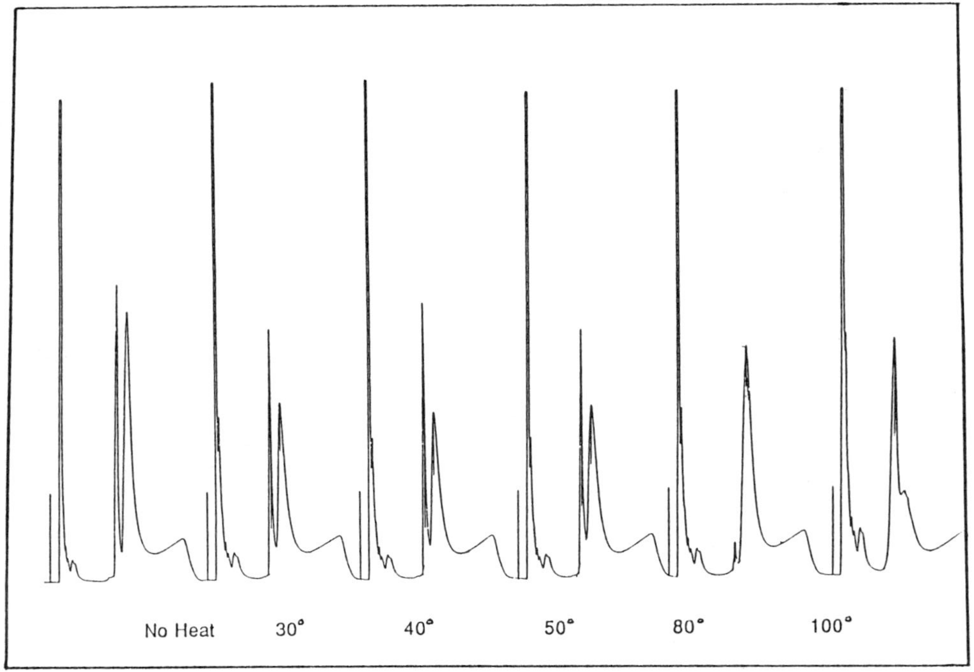

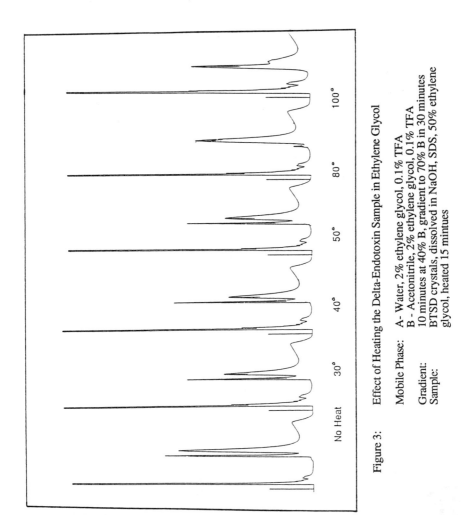

Figure 3: Effect of Heating the Delta-Endotoxin Sample in Ethylene Glycol

Mobile Phase: A- Water, 2% ethylene glycol, 0.1% TFA
B - Acetonitrile, 2% ethylene glycol, 0.1% TFA
Gradient: 10 minutes at 40% B, gradient to 70% B in 30 minutes
Sample: BTSD crystals, dissolved in NaOH, SDS, 50% ethylene glycol, heated 15 mintues

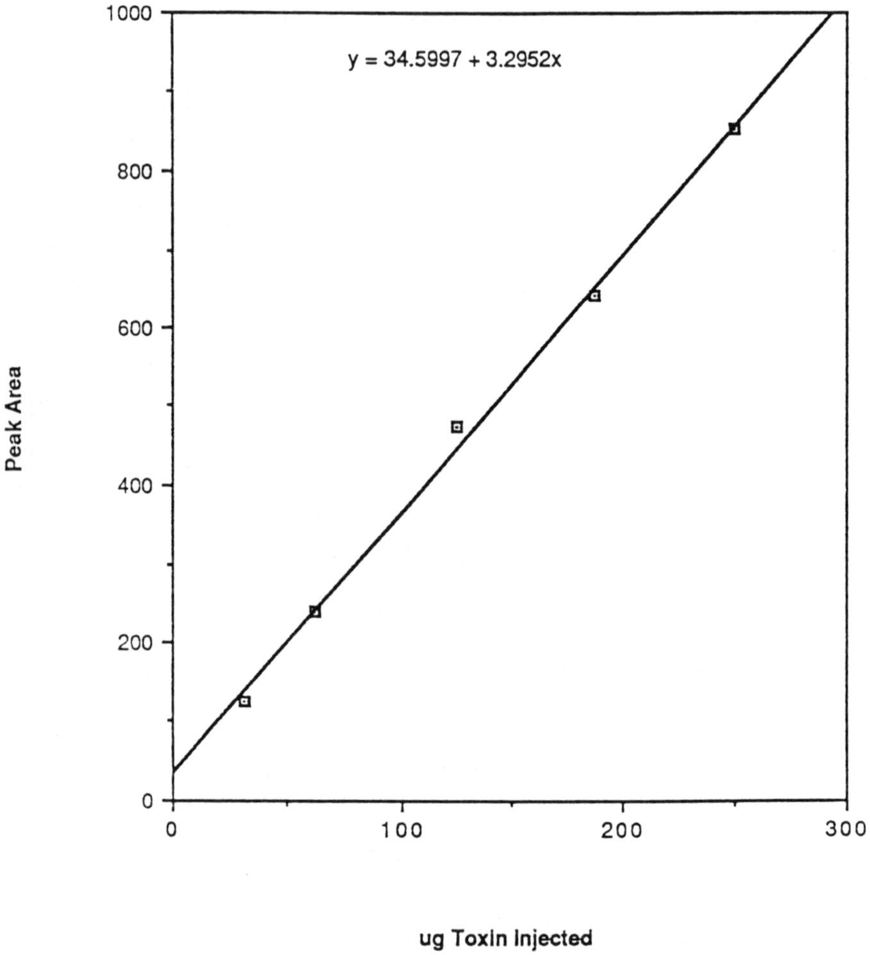

Figure 4. *Bacillus thuringiensis* var. *san diego* delta-endotoxin Calibration Curve

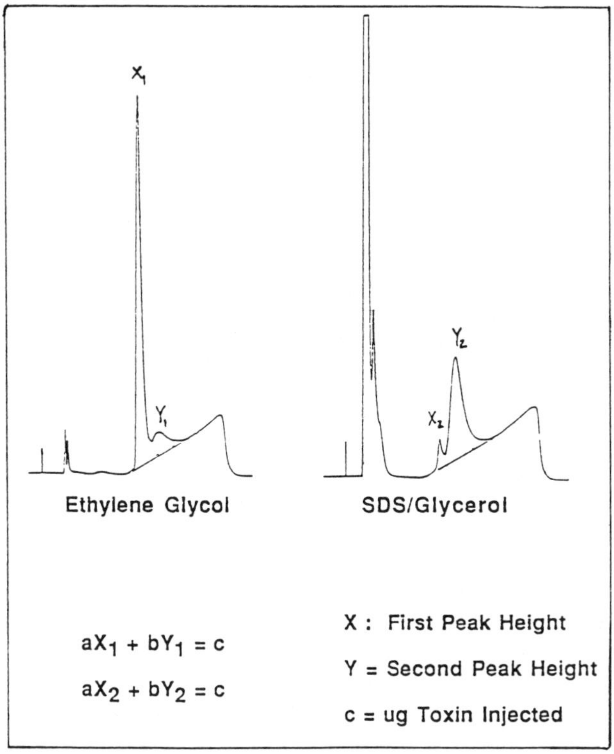

Figure 5. Quantitation of BTSD toxin

phase. Second is the effect of 50% ethylene glycol and heat on the retention time. Based on the effect of added SDS, it appears the glycol stabilizes the toxin in a relatively native state.

Literature Cited

1. Edwards, D. L.; Herrnstadt, C; Soares, G. G.; Wilcox, E. R. Bio/Technology 1986, 4, 305-308.
2. Bennet, B. D.; Gaertner, F. H.; Gilroy, T. E.; Herrnstadt, C.; Sobieski, D. A. Gene 1987, 57, 37-46.
3. Andrews, A. T. Electrophoresis, Theory, Techniques and Biochemical and Clinical Applications; 2nd Edition: Clarendon Press Oxford, 1986.
4. Private communication, Sept. 1989, Pat Nolan, Mycogen Corp.

RECEIVED March 5, 1990

Chapter 9

Use of Sodium Dodecyl Sulfate–Polyacrylamide Gel Electrophoresis To Quantify *Bacillus thuringiensis* δ-Endotoxins

Susan M. Brussock and Thomas C. Currier[1]

Ecogen Inc., 2005 Cabot Boulevard West, Langhorne, PA 19047–1810

A reliable method for measuring the amount of *Bacillus thuringiensis* (BT) crystalline delta (δ)-endotoxin proteins in production samples has been developed. This method consists of inhibiting BT proteases by high pH treatment prior to separating δ-endotoxin proteins from growth medium proteins by sodium dodecyl sulfate-polyacrylamide gel electrophoresis (SDS-PAGE). Toxin protein bands are then quantified by densitometry. In principle, any toxin protein can be quantified, providing it can be solubilized with little breakdown, and a standard curve can be generated from a purified sample of the toxin protein to be measured. Toxin proteins produced by diverse BT strains grown in various fermentation media have been quantified by this method, as have δ-endotoxins in different types of commercial formulations.

Bacillus thuringiensis (BT) is characterized by its ability to produce crystalline proteins known as δ-endotoxins. Several types of δ-endotoxins (toxin proteins) exhibiting different insecticidal activities have been identified, including CryIA and CryII toxin proteins, both active against lepidopterans, and the coleopteran-active CryIIIA protein (see 1 for descriptions and details of these and other δ-endotoxins). Commercial BT bioinsecticide products have been used in the United States since 1958 to control pests of crops and forests. The highly specific insecticidal activity, lack of mammalian, avian, and beneficial insect toxicity, as well as the environmentally innocuous nature of δ-endotoxins have stimulated renewed interests in BT products.

Quantification of BT products has traditionally been in terms of International Units (IU's), based on a product's insect bioassay activity

[1]Current address: Sterling Drug Inc., 25 Great Valley Parkway, Malvern, PA 19355

0097–6156/90/0432–0078$06.00/0

compared to that of a standard. Recent changes in the United States Environmental Protection Agency (EPA) BT Registration Standard require that, by the end of 1990, all BT product labels list the amount of toxin protein as percent (by weight) active ingredient (% a.i.) (2); IU's will no longer be the accepted method of measurement. BT products containing a mixture of δ-endotoxin classes active against different insect groups (for example, lepidopteran and coleopteran) will have to specify the % a.i. of each type of toxin.

Fermentation media typically used for BT production and the nature of BT itself make the quantification of % a.i. quite complex. Many cost-effective production media contain large amounts of crude, insoluble proteins which must be distinguished from the toxin proteins to be quantified. We have selected the common laboratory technique of sodium dodecyl sulfate-polyacrylamide gel electrophoresis (SDS-PAGE) to resolve toxin proteins from growth medium proteins. BT grown in crude media, however, produce high levels of protease that will degrade toxin proteins as the toxin crystals are solubilized for electrophoresis. We have found that inhibition of BT production sample proteases by treatments with a variety of protease inhibitors such as leupeptin, pepstatin, EDTA (ethylene-diamine-tetraacetic acid), and PMSF (phenylmethylsulfonyl fluoride) is not accomplished in a workable time frame (data not shown). The most practical way we have found to inhibit proteolytic activity in these samples is high pH treatment before crystal toxin solubilization. Figure 1 demonstrates the increase in the amount of toxin proteins, especially CryIA (approximately 131 to 133kDa), that appears on gels after high pH treatment of BT production samples. The amount of toxin protein measured in a given sample after high pH treatment is reproducible and consistent (data not shown).

Described herein is a quantification method for BT toxin proteins (PGel Assay). We have used this PGel Assay to quantify BT toxin proteins in a variety of production samples ranging from fermentation broths to technical powders and formulated end products. These samples include fermentation broths and concentrates; spray dried, freeze dried, and acetone powders; and oil flowable, aqueous flowable, and wettable powder formulations. This PGel Assay is reliable, reproducible, and relatively easy to perform. (See Groat, *et al.*, this volume, for a description of a quantitative immunoassay for CryIA toxin protein in BT production samples.)

PGel Assay for BT Toxin Proteins

Overview. The first step in this assay involves treatment of the BT sample to be quantified with high pH to inhibit proteases. High pH treatment partially, though not completely, solubilizes BT toxin protein crystals. Crystal solubilization and protein denaturation is completed by high temperature incubation in Laemmli sample buffer (3). Multiple sample inputs of the unknown sample and of a standard purified crystal preparation are resolved by SDS-PAGE. Protein bands are visualized by staining with Coomassie Brilliant Blue and specific toxin protein bands are quantified by densitometry. A standard curve is generated from measured band volumes and the toxin amounts in samples are determined by interpolation.

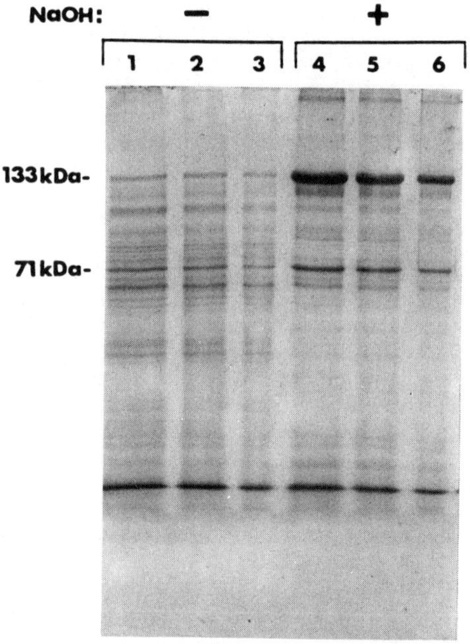

Figure 1. SDS-PAGE of a BT spray dried powder containing CryIA (133 kDa) and CryII (71 kDa) toxin proteins. The same BT powder was run in all lanes. Lanes 4-6 contain three different loading volumes of the powder sample treated with high pH to inhibit proteases by the addition of NaOH, then neutralized with HEPES buffer prior to SDS-PAGE. Lanes 1-3 contain the same loading volumes of the powder sample treated with water in place of NaOH, then HEPES buffer. Note the increased amounts of CryIA and CryII toxin proteins that appear after high pH treatment.

SDS-PAGE. We use a standard discontinuous SDS-PAGE system (3) with a 10% acrylamide resolving gel. Linear gradient resolving gels from 5% to 20% acrylamide have also been used. Electrophoresis is conveniently done using a Hoefer Demi-Gel rig (Hoefer Scientific Instruments, San Francisco, CA) with 0.75 mm spacers and a 20-well comb, according to the manufacturer's instructions. Protein bands are visualized by staining with 0.2% Coomassie Brilliant Blue R in 50% methanol and 10% acetic acid, usually overnight. Destaining is done in 25% methanol and 10% acetic acid until the gel background is clear.

Standards. Standards are necessary to relate the data obtained from a densitometric scan of the sample toxin bands to a known quantity of toxin. The Coomassie dye used to visualize the protein bands gives a linear response to amount of protein. However, this Coomassie dye binds to different toxin proteins to varying extents (see Figure 2), making it

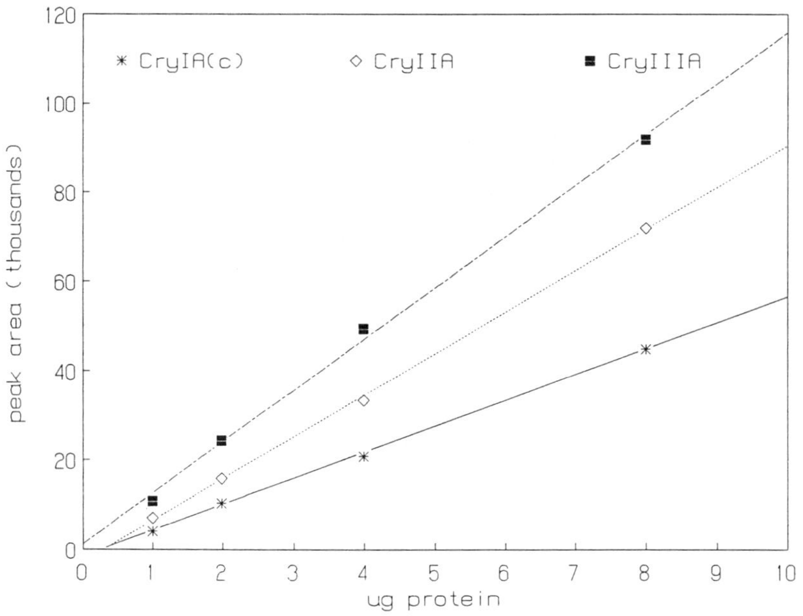

Figure 2. Example of differences in Coomassie dye binding and the linear working range of purified CryIA, CryIIA, and CryIIIA toxin protein bands quantified by densitometry after SDS-PAGE. Coomassie dye binding is linear to at least 10 ug/band.

necessary to use a separate standard for each type of δ-endotoxin to be quantified. For example, to quantify the amount of CryIA in a sample, purified CryIA toxin protein is used as a standard. Most crystalline δ-endotoxins are readily purified by density gradient centrifugation (4-8) or by solubilization and recrystallization (9,10). Purified toxin proteins are stored in crystalline form at -20° C in water containing 0.005% Triton X-100 (added to prevent purified crystals from sticking to surfaces and to each other).

Chemical assays for total protein, such as the Bio-Rad Protein Assay (Bio-Rad Laboratories, Richmond, CA) or the BCA Protein Assay Reagent (Pierce, Rockford, IL), are used to determine the concentration of toxin in the purified crystal stock. Known concentrations of purified toxin protein are loaded in each lane. We have determined that the optimum working range of the standard curve is between 0.5 ug and 10 ug for CryIA, CryIIA, and CryIIIA toxin proteins (see Figure 2). Due to gel staining and destaining variations, a set of standards is included in each gel.

Sample Preparation. Samples are diluted with water as necessary to give amounts of toxin protein loaded on the gel that are within the linear range of the standard curve. The diluted sample is sonicated and a subsample is removed to a microfuge tube. The aliquot is briefly treated at high pH with 0.1N NaOH (final concentration) to inactivate proteases, then neutralized with 0.1x volume of 3M HEPES (N-2-hydroxyethylpiperazine-N'-2-ethanesulfonic acid). A small volume of the neutralized sample is placed in a fresh microfuge tube, 0.5x volume of 3x Laemmli sample buffer is added, and the mixture is heated at 100° C for 5 minutes. Before loading, each sample is centrifuged for 5 minutes in a microfuge to pellet insolubles. Purified toxin protein for standard curve generation is treated only with Laemmli sample buffer and heated for 5 minutes at 100° C. We have found that treatment of standards with high pH is unnecessary because proteases have been removed during the purification process. As Figure 3 demonstrates, we find no difference in Coomassie dye binding between toxin protein standards that have been treated with NaOH and those that have not.

Loading of processed samples onto the gel is perhaps the most difficult technical aspect of this assay. Great care must be taken to accurately measure the volumes loaded of both samples and standards. Standards are loaded at exact protein amounts, the volumes depending on

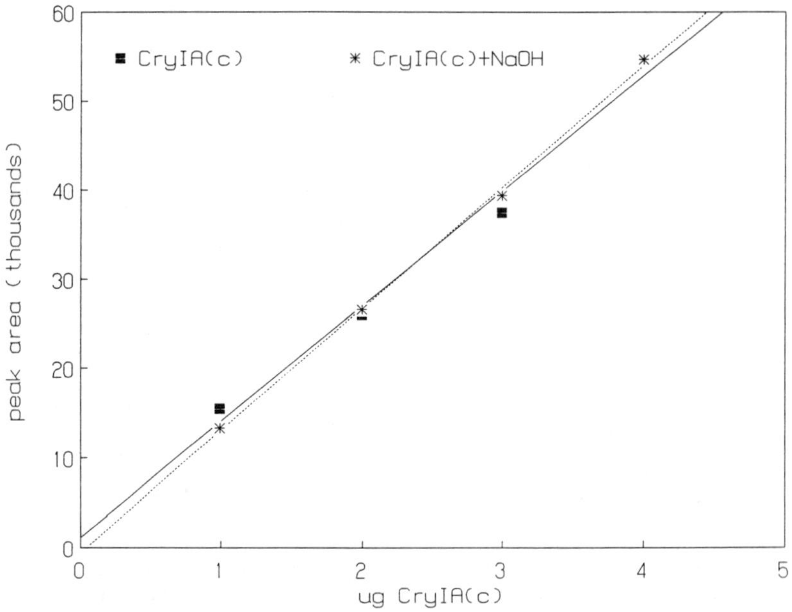

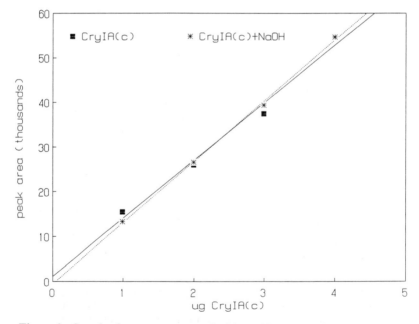

Figure 3. Standard curves generated with purified CryIA(c) toxin protein treated with and without NaOH + HEPES prior to SDS-PAGE. Peak areas were determined by a densitometric scan of the CryIA(c) toxin protein bands.

the protein concentration of the standard stock plus the dilution made by the addition of Laemmli sample buffer. A typical standard curve consists of four points: 4 ug, 3 ug, 2 ug, and 1 ug. Samples are loaded at small volumes (≤ 10 ul) to minimize protein band spreading caused by high salt concentration in the treated samples.

Figure 4 shows an example of SDS-PAGE to quantify CryIA toxin proteins in various samples. Lanes 1-4 contain purified CryIA toxin protein standards at 4 ug - 1 ug, while lanes 5-19 contain the technical powder and formulated samples to be quantified (3 lanes per sample). Each of these samples also contains CryII toxin protein, molecular weight of approximately 71 kDa, not quantified on this gel. A separate gel containing purified CryII toxin protein standards would be run to determine the amount of CryII toxin protein in these samples. Growth medium proteins, due to size heterogeneity, do not show distinct protein bands that interfere with toxin protein bands. After staining and destaining, a gel such as the one shown in Figure 4 is ready to be scanned on a densitometer for toxin protein quantification.

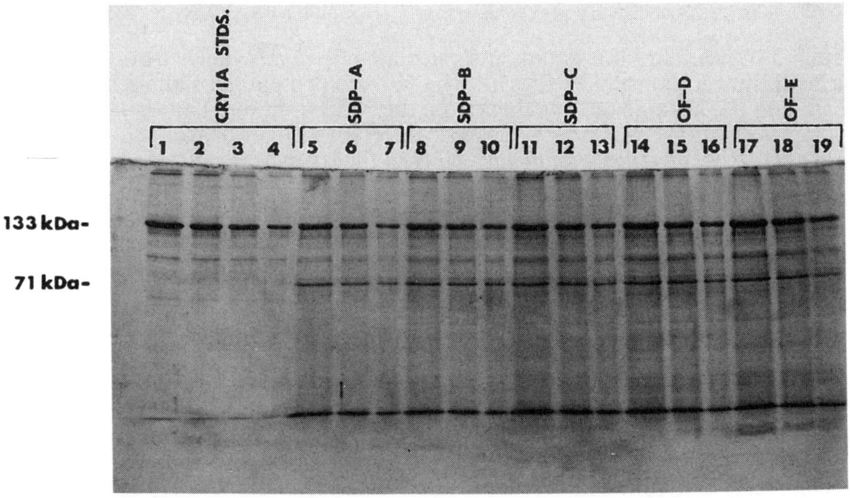

Figure 4. Example of a PGel Assay gel used to quantify the amount of CryIA toxin protein (133 kDa) in various BT samples. Lanes 1-4 contain purified CryIA toxin protein standards. Lanes 5-7, 8-10, and 11-13 contain three different spray dried powders (SDP-A - SDP-C). Lanes 14-16 and 17-19 contain two different oil flowable formulations (OF-D and OF-E). Each sample was loaded at three different volumes in three wells. Each sample also contains CryII toxin protein (71 kDa), not quantified on this gel. Densitometric quantifications of the CryIA toxin bands and % CryIA determinations for each sample are shown in Table I.

Densitometric Scanning. Gels are typically scanned wet, immediately after destaining, but they can also be scanned after drying between two sheets of cellophane (available from Bio-Rad Laboratories). The nonuniform nature of bands on a protein gel requires that the entire toxin band be scanned using either a ribbon densitometer or one capable of a two-dimensional scan (a point source scan is not adequate). The densitometric quantification is expressed as peak area or peak volume per band. This data is used to generate a standard curve and to calculate sample toxin values.

Calculations. A standard curve is constructed from the densitometric quantification of the known inputs toxin protein. For samples, the amount of toxin per band is interpolated from the standard curve and the % toxin protein per sample is calculated using the following equation:

$$\%\ \text{toxin protein} = \frac{\text{ug toxin from standard curve}}{\text{ul sample loaded}} \times \frac{\text{dilution factor}}{\text{ug/ul concentration of original stock}} \times 100\%$$

Table I shows scan information and calculated % CryIA values from a laser scan of the gel shown in Figure 4. The % toxin protein per sample is based on the average of the values from the 3 different input amounts. Samples should be run several times on different gels for proper statistical analysis. Recently, the overall variability of this assay was estimated to be approximately 8.5%.

Applications of PGel Assay Data

We currently use this PGel Assay to determine the % a.i. of BT products in compliance with the new EPA labeling regulations. Along with this important function, the data obtained from the PGel Assay have other applications. The % a.i. of a sample can be used in bioprocess methods evaluations, (e.g., fermentation media development and downstream processing). In addition, toxin protein quantification can be used in conjunction with insect bioassay to determine specific activities and relative potencies of newly created or discovered δ-endotoxins. For some samples, PGel Assay data (although certainly not a measure of bioactivity) does correlate with bioassay data. In these instances, the PGel Assay can be used as a screening tool to quickly screen samples for the more time-consuming insect bioassay. Two examples of correlations we have found between bioassay and PGel Assay data are presented in Tables II and III. Table II lists LC_{50}'s of spray dried powder 'G' and subsequent formulations of this powder, as determined by diet incorporation bioassays against 6 day old *Plutella xylostella* larvae. The PGel Assay was used to determine the % a.i. of each sample (% CryIA plus % CryII). When the LC_{50} determinations are adjusted to reflect the % toxin protein per sample (pLC_{50}) the resulting values are virtually identical. This is expected, since the same powder was used in each formulation. Linear regression analysis of % a.i. versus LC_{50} per sample shows a high correlation (R^2 =0.851).

Table I. Example of a typical densitometric scan report and calculated % CryIA[a] of various BT samples

Lane	Sample[b]	Input	Peak Volume	Standard Curve	
1	CryIA	4ug	402504		
2	CryIA	3ug	295676	y = 100196x +/- 317	
3	CryIA	2ug	206962	R^2 = 0.999	
4	CryIA	1ug	98088		
				%CRYIA[c]	mean %CRYIA +/- S.D.
5	SDP-A	6ul	195290	8.03	
6	SDP-A	4ul	127287	7.84	7.88 +/- 0.13
7	SDP-A	2ul	63220	7.77	
8	SDP-B	6ul	216100	8.88	
9	SDP-B	4ul	147208	9.07	9.16 +/- 0.33
10	SDP-B	2ul	77399	9.52	
11	SDP-C	6ul	259420	10.67	
12	SDP-C	4ul	174903	10.78	10.73 +/- 0.06
13	SDP-C	2ul	87191	10.73	
14	OF-D	6ul	276142	2.37	
15	OF-D	4ul	174864	2.25	2.35 +/- 0.09
16	OF-D	2ul	94638	2.43	
17	OF-E	6ul	378569	3.24	
18	OF-E	4ul	252020	3.24	3.22 +/- 0.04
19	OF-E	2ul	123660	3.17	

[a]The scan report detailed above is from a densitometric scan (300A Computing Densitometer, Molecular Dynamics, Sunnyvale, CA) of the CryIA toxin protein bands present on the gel shown in Figure 4.
[b]SDP = spray dried powder, OF = oil flowable formulation
[c]For formula, refer to Calculations.

Table II. *Plutella xylostella* insecticidal activity and % active ingredient as determined by PGel Assay of a BT technical powder and three formulations of that powder

Sample[a]	% a.i.	mean LC_{50}[b] ug sample/ml diet	mean pLC_{50} ug toxin/ml diet
SDP-G	24.1	0.34[c]	0.08
WP	12.5	0.7	0.09
OF	5.8	1.2	0.07
AF	3.3	1.8	0.06

[a]SDP = spray dried powder, WP = wettable powder, OF = oil flowable, AF = aqueous flowable
[b]Mean of two 8 dose bioassays, 30 insects per dose.
[c]One 8 dose determination only.

Another example of correlation between insecticidal activity and % a.i. is given in Table III. In this example, the LC_{50} values of five different production powders (H-L) of the same BT strain were determined by diet surface contamination bioassay against *Heliothis virescens* neonates; % CryIA was determined by PGel Assay. Normalizing these LC_{50} values with respect to % CryIA protein per surface area of diet (pLC_{50}) yields toxin protein concentrations that are similar. Linear regression analysis of % CryIA protein versus LC_{50} of the entire sample gives a high correlation (R^2 = 0.885).

Table III. *Heliothis virescens* insecticidal activity and % CryIA as determined by PGel Assay of five technical powders of the same BT strain

Sample[a]	% CryIA	mean LC_{50}[b] ng sample/600mm² diet	mean pLC_{50} CryIA/600mm² diet
SDP-H	21.7	19.0[c]	4.1
SDP-I	21.2	15.5	3.3
SDP-J	17.9	20.5	3.7
SDP-K	13.0	28.0	3.6
SDP-L	10.3	40.5	4.2

[a]SDP = spray dried powder
[b]Mean of two 8 dose bioassays/30 insects per dose
[c]One 8 dose determination only

Summary

The PGel Assay described here reproducibly quantifies the amount of BT δ-endotoxins present in production samples ranging from fermentation broths to various commercial formulations. This assay is versatile in that a wide range of sample types can be handled without significant procedural changes, and a variety of different toxin proteins can be quantified. Results obtained using this assay are currently used to meet EPA labeling requirements, and can aid in bioprocess development evaluations and novel BT toxin protein evaluations. In some instances, PGel Assay data do correlate with bioassay data. This PGel Assay is relatively quick, easy to perform and, with the exception of a ribbon or two-dimensional densitometer, does not require specialized reagents or equipment.

Acknowledgments

We thank Marion Burgwin and Tim Johnson for bioassay data and Catrina Banks and Yvonne Gattelli for PGel Assay data.

Literature Cited

1. Höfte, H.; Whiteley, H. R. Microbiol. Reviews 1989, 53(2),242-55.
2. Registration Standard for the Reregistration of Pesticide Products Containing *Bacillus thuringiensis* as the Active Ingredient, Case #0247, U.S. Environmental Protection Agency, Office of Pesticide Programs: Washington D.C., 1988.
3. Laemmli, U. K. Nature 1970, 227,680-685.
4. Fitz-James, P. C.; Gillespie, J. B.; Loewy, D. J. Invert. Path. 1984, 43, 47-58.
5. Ang, B. J.; Nickerson, K. W. Appl. Envir. Microbiol. 1978, 36,625-26.
6. Li, R. S.; Jarrett, P.; Burges, H. D. J. Invert. Path.1987, 50,277-84.
7. Herrnstadt, C.; Soars, G. G.; Wilcox, E. R.; Edwards, D. L. Bio/tech. 1986, 4,305-08.
8. Zhu, Y. S.; Brookes, A.; Carlson, K.; Filner, P. Appl. Envir. Microbiol. 1989, 55(5),1279-81.
9. Garfield, J. L.; Stout, C. D. J. Biol. Chem. 1988 263(24),11800-801.
10. Li, J.; Henderson, R.; Carroll, J.; Ellar, D. J. Mol.Biol. 1988, 199, 543-44.

RECEIVED February 19, 1990

Chapter 10

Quantitative Immunoassay of Insecticidal Proteins in *Bacillus thuringiensis* Products

R. Gene Groat, James W. Mattison, and Eric J. French

Ecogen Inc., 2005 Cabot Boulevard West, Langhorne, PA 19047–1810

We are applying microplate ELISA technology for quantitative immunoassay of insecticidal (toxin) proteins in *Bacillus thuringiensis* (BT) production samples. Once perfected, this assay format is fast, accurate, and readily adapted to automation. High protease levels and multiple, related toxin proteins in BT production samples require carefully constructed primary antibody reagents and assay conditions for ELISA. For immunoassay of CryIA lepidopteran active toxin proteins, we have developed a synthetic peptide directed, site specific primary antibody reagent that reacts equally with individual CryIA(a), CryIA(b), and CryIA(c) toxin proteins. Using this antibody reagent, total CryIA toxin protein is accurately determined in BT production samples of a variety of strains in diverse fermentation media. By a similar approach, we have also developed primary antibody reagents that distinguish individual CryIA toxin proteins (80-90% identical at the amino acid sequence level) in mixed samples. Data from these immunoassays are corroborated by data from other chemical and physicochemical methods.

The number of commercial formulations of *Bacillus thuringiensis* (BT) based biopesticide products is rapidly increasing. The primary active ingredients in these products are insecticidal proteins produced in BT during sporulation as crystalline inclusions that are released along with spores upon lysis of sporangia. Details of the genetics, nomenclature, specificity, mode of action, and applications of BT crystal (toxin) proteins are most recently reviewed by Hofte and Whiteley (1). Historically, the active ingredients in commercial BT products have been described in relative 'International Units' of insecticidal activity determined under standard bioassay conditions using a single insect species, relative to an 'International Standard' BT sample. This 'International Units' description is no longer suitable as standard bioassays and reference samples increasingly lag behind development of new BT products with different toxin proteins and target applications. The United States Environmental

0097–6156/90/0432–0088$06.00/0

Protection Agency (EPA) has recently required (2) that all commercial BT products be labelled for %(w/w) active ingredient toxin proteins by class (insect order specificity).

Beyond such regulatory requirements are industrial applications where quantitation of BT toxin proteins is desirable. Insect bioassay data (*eg* LC50, ng sample per unit diet) combined with toxin protein assay data (*eg* %(w/w) lepidopteran active toxin protein(s)) gives a measure of specific activity (*eg* PLC50, insecticidal activity per unit toxin protein) that allows for comparisons of the relative potencies of different BT toxin proteins or samples. New BT strain evaluations or bioprocess development and quality control programs for standard BT strains are greatly facilitated by specific activity data. In production settings, insect bioassays suffer from high variability, space and personnel requirements, slow completion, and low sample throughput. Specific biochemical assays for BT toxin proteins can be more precise, faster to complete, and have higher sample throughput.

Several enzyme linked immunosorbant assays (ELISAs) have been reported for detection or quantitation of BT *kurstaki* and BT *israelensis* toxin proteins (3-7). We find that these ELISA protocols are not suitable for BT production samples containing high levels of protease activity. Significant protease activity in samples can result in breakdown of BT toxin proteins during sample preparation and primary antibody incubation steps, giving erroneous results. Brussock and Currier report in this volume on a physicochemical (gel electrophoresis) method for quantitative assay of specific BT toxin protein classes in production samples. Good correlation between data from this assay and insect bioassay data has been demonstrated, and such quantitative protein gel electrophoresis (protein gel) assays are used routinely at Ecogen. We have also developed immunochemical methods for quantitative assay of specific BT toxin protein classes in production samples. This paper describes a quantitative immunoassay for determination of total BT CryIA toxin proteins (see 1 for nomenclature) in production samples using microplate ELISA technology. Quantitative immunoassay of CryIA lepidopteran active toxin proteins for naturally occuring BT strains presents special problems because varying amounts of multiple, related CryIA toxin proteins (80-90% identical at the amino acid sequence level) are typically present.

BT Production Samples

Development of quantitative assays for BT toxin proteins in production samples is complicated by the presence of significant and variable amounts of heterologous media proteins and BT derived proteases. Chemical assays for total protein are not suitable for assays of BT toxin proteins in production samples because of the presence of heterologous media proteins. In protein gel assays this problem is solved by physically separating toxin proteins from heterologous proteins using electrophoresis. In immunoassays this problem is solved by using specific antibody reagents that biochemically discriminate toxin proteins in the presence of heterologous proteins.

In commercial production, BT is frequently grown on crude soluble and insoluble media proteins as a primary source of nitrogen and/or carbon. In effect, BT is selected for high protease production in order to utilize the crude media protein for growth, and final fermentation products can contain high levels of protease activity. Significant amounts of these BT derived proteases can remain even in technical powders and formulations. In crystalline form, BT toxin proteins are relatively resistant to these proteases, but once crytalline toxin proteins are solubilized for assay they can be rapidly degraded. Protease activities in solubilized samples can be a major problem for any quantitative assay for BT toxin proteins in production samples.

Theoretical and Practical Advantages of ELISA

The biological specificity of antibody reagents circumvents the need for physical separation of specific proteins in complex mixtures. This minimizes sample preparation steps for specific protein immunoassays. In practical applications, a major advantage of immunoassays in general and microplate ELISAs in particular is high sample throughput. The 8 by 12 well microplate format allows for 96 individual tests to be handled in a concerted manner in all but the initial assay steps. Furthermore, technologies for automation of microplate ELISAs are highly developed and commercially available. Compared to protein gel assay formats, an order of magnitude or more advantage is easily obtained for sample throughput in microplate ELISA formats. Sensitivities of ELISAs are also theoretically much higher than protein gel assays, but this is not a significant consideration here, since levels of BT toxin protein analytes in production samples are high. A disadvantage of immunoassays compared to protein gel assays for BT toxin proteins is the requirement for specialized primary antibody reagents.

Site Directed, Synthetic Peptide Antibody Reagents - Advantages for CryIA ELISAs

First generation commercial BT products are formulations of BT *kurstaki* strains (8). Such products contain variable amounts of 3 related lepidopteran active CryIA toxin proteins. Complete amino acid sequences of individual CryIA(a), CryIA(b), and CryIA(c) toxin proteins are known from translation of DNA sequence information and these 3 CryIA toxin proteins are 80-90% identical at the amino acid sequence level (see 1 for details of CryIA sequence homologies). Most polyclonal antibodies and monoclonal antibodies raised to purified CryIA toxin proteins will react to different extents with individual CryIA toxin proteins. Such antibodies are not suitable for precise, quantitative immunoassay of total CryIA toxin proteins in mixed samples, since relative proportions of the 3 CryIA toxin proteins can vary.

For quantitative immunoassay of total CryIA(a), CryIA(b), or CryIA(c) toxin proteins, we have developed a synthetic peptide directed, site specific primary antibody reagent that reacts equally with each of these 3 CryIA toxin proteins. This antibody reagent was raised to a 12-mer synthetic peptide representing a conserved, N terminal amino acid sequence that is identical in CryIA(a), CryIA(b), and CryIA(c) toxin proteins. An additional advantage of our total CryIA synthetic peptide antibody reagent is that it reacts stoichiometrically with a small, protease resistant piece of these proteins (unpublished results). Therefore, our total CryIA ELISA is relatively tolerant of proteolytic breakdown of CryIA toxin proteins in the course of assay.

By a similar approach, using synthetic peptides representing unique amino acid sequences in CryIA(a), CryIA(b), and CryIA(c) proteins, we have also developed primary antibody reagents that distinguish individual CryIA toxin proteins in mixed samples. CryIA protein directed monoclonal antibody approaches to this problem are not always successful. For monoclonal antibodies raised to purified toxin proteins, the immunized animal or spleen 'decides' which sites (epitopes) of the protein to direct antibodies toward. For these highly conserved CryIA toxin proteins, such immunological hot spots appear to dominate monoclonal antibody responses to whole protein immunogens to such an extent that single monoclonal antibodies capable of distinguishing individual CryIA toxin proteins are not always obtained (9). In synthetic peptide immunogen approaches, the researcher 'decides' at least the potential specificity of antibodies obtained and one can direct antibody specificity toward relatively rare unique amino acid sequences in CryIA(a), CryIA(b), or CryIA(c) toxin proteins.

Specificity of CryIA Derived Synthetic Peptide Antibody Reagents

Figure 1 shows qualitative specificities of total CryIA and specific CryIA(c) synthetic peptide antibody reagents in Western blot analysis (10). Using a synthetic peptide approach, we have developed primary antibody reagents that react equally with CryIA(a), CryIA(b), and CryIA(c) toxin proteins (Figure 1A), or specifically with only CryIA(c) toxin protein (Figure 1B). The samples analyzed in Figure 1 are density gradient purified CryIA(a), CryIA(b), and CryIA(c) crystals. High sensitivity immunochemical (alkaline phosphatase linked) and chemical (colloidal gold/silver) staining methods detect many minor breakdown products of the 131 to 133 kDa CryIA toxin proteins, probably generated in the course of sample preparation for Western blot analysis. Profiles of these breakdown products are somewhat different for individual CryIA proteins (see Figure 1A). Although these proteins are 80% to 90% identical at the amino acid sequence level, chemical colloidal gold/silver staining properties of CryIA(a), CryIA(b), and CryIA(c) proteins are quite different (Figure 1C). Quantitatively, by ELISA, the specificities of these synthetic peptide antibody reagents are verified (Figure 2A and B, respectively).

Total CryIA ELISA protocol

A total CryIA ELISA protocol is outlined below:

1. Weigh out samples/standards, solubilize toxin protein, make appropriate dilutions.
2. Input sample/standard aliquots to microplate wells, incubate.
3. Wash wells.
4. React wells with primary antibody - CryIA synthetic peptide antibody (mouse IgG).
5. Wash wells.
6. React wells with enzyme linked second antibody - *eg* alkaline phosphatase linked goat anti-(mouse IgG).
7. Wash wells.
8. Develop enzyme reaction (alkaline phosphatase) in the wells.
9. Quantitate the final assay signal (absorbance) in a microplate reader.
10. Calculate sample values from the standard curve.

This is a relatively simple and straightforward ELISA protocol. High abundance of analyte in BT production samples allows for direct, nonspecific binding of CryIA toxin proteins to the microplate solid phase (plastic) as an initial step in this assay. Subsequent assay steps are quite standard and, except for primary antibody reagent, use commercially available materials. Washes and antibody incubations are done in a Tris buffered saline solution, pH 7.8.

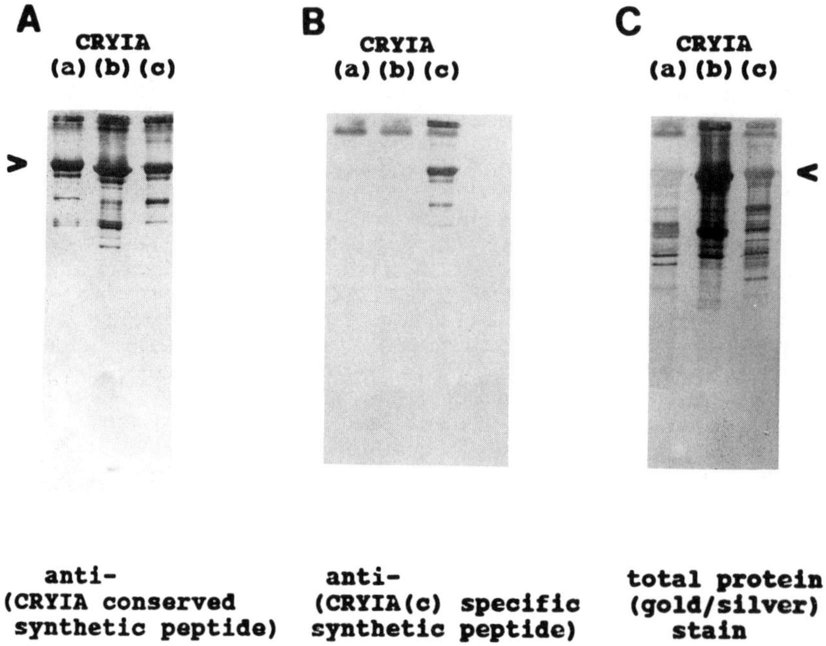

Figure 1. Specificities of CryIA derived synthetic peptide antibody reagents in Western blot analysis. Replicate samples of density gradient purified CryIA(a), CryIA(b), and CryIA(c) toxin protein crystals (approximately 2 ug per lane) were subjected to sodium dodecyl sulfate polyacrylamide gel electrophoresis, transferred to nitrocellulose, and probed with CryIA derived synthetic peptide antibody reagents. Arrows indicate full length (131 to 133 kDa) CryIA toxin proteins. (Methyl Green dye bands mark the top of each lane.)

A: Antibody probe = anti-(CryIA conserved synthetic peptide)
B: Antibody probe = anti-(CryIA(c) unique synthetic peptide)
C: Total protein staining (colloidal gold/silver)

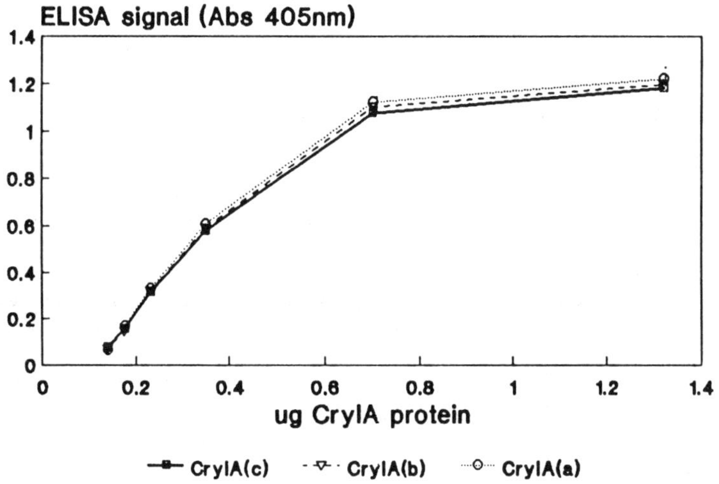

A. CryIA conserved peptide

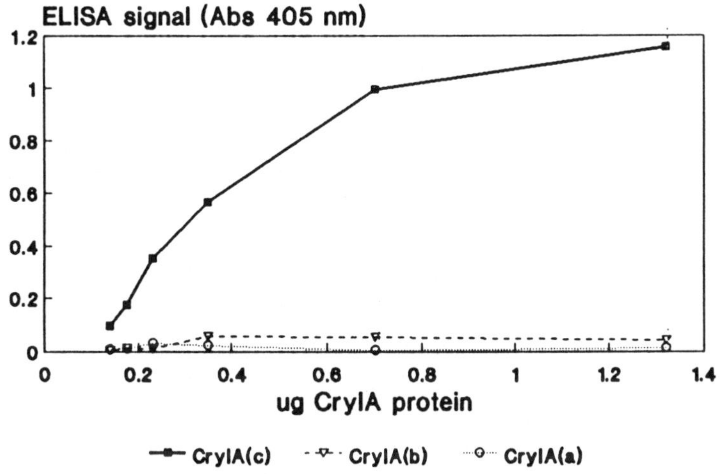

B. CryIA(c) specific peptide

Figure 2. Specificities of CryIA derived synthetic peptide antibody reagents in ELISA. Density gradient purified CryIA(a), CryIA(b), or CryIA(c) toxin protein crystals were solubilized, quantitated by chemical protein assay, and dilution series were analyzed by ELISA.

A: Primary antibody = anti-(CryIA conserved synthetic peptide)

B: Primary antibody = anti-(CryIA(c) unique synthetic peptide)

Total CryIA ELISA Data Analysis

Figure 3 shows a typical standard curve for our total CryIA ELISA. As for protein gel assays (see Brussock and Currier, this volume), standards are density gradient purified crystal preparations that have been quantitated by a chemical assay for total protein. The optimum working range for our total CryIA ELISA is about 0.1 to 0.8 ug CryIA toxin protein per well. The working range of this assay can be manipulated somewhat, using different types of microplates (plastics) or different specific conditions for individual assay steps (*eg* analyte binding, dilutions of primary and secondary antibody reagents, buffer conditions for antibody incubation and wash steps). As mentioned earlier, assay sensitivity is not an issue in this application. In fact, dilutions into our working range are large for these samples. Dose-response curves for the working range of our assay are quite linear, but are slightly better fit by a second order polynomial (see Figure 3). Computer software is used for standard curve fitting and calculations of sample values by interpolation. A separate internal standard curve is included in each microplate to correct for plate to plate or day to day differences in assay performance. Multiple weighouts of each sample and multiple inputs of each weighout are routinely included in each plate, but within-plate variability (typically less than 5%) is much less than between-plate variability (about 10%) for this assay. It is therefore necessary to measure any given sample 3 or more times in separate assays (plates) for proper statistical analysis of assay results.

Total CryIA ELISA Assay Results

Our results of total CryIA ELISA and protein gel assay (PGel Assay; Brussock and Currier, this volume) for BT production samples compare well. Table 1 shows results of both of these assays for a BT production sample (technical powder). This sample was measured 20 times by each method in separate assays (plates or gels). Within-assay replicates were higher for ELISA (N = 12) than for PGel Assay (N = 3), reflecting higher sample throughput for ELISA. The grand mean %(w/w) CryIA toxin protein values for this technical powder sample determined by this ELISA and PGel Assay were similar (21.18% and 19.57%, respectively, see Table 1). Note that, for this ELISA or for this PGel Assay, between-assay standard deviations dominate within-assay standard deviations. We find that mean values for these assays are normally distributed and standard deviations are distributed as Chi-squared.

Generally, for a wide range of BT production samples, we find that CryIA toxin values determined by ELISA or by PGel Assay are in good agreement. The nature of these 2 methods is fundamentally different (one immunochemical and one physicochemical); therefore, agreement of results from CryIA ELISA and PGel Assay indicates that these experimentally determined toxin protein values are valid. Experimentally determined toxin protein values for crude media BT production samples are potentially underestimates of true mean values, since protease activities in these samples can be high. For some samples, mean CryIA ELISA values are slightly higher than mean PGel Assay values, reflecting this ELISA's relative insensitivity to proteolytic breakdown of CryIA toxin protein in the course of assay. Statistical analyses of recent CryIA ELISA and CryIA PGel Assay results for a variety of samples show that overall assay variability for CryIA ELISAs (12.5%) is somewhat higher than for CryIA PGel Assays (8.5%).

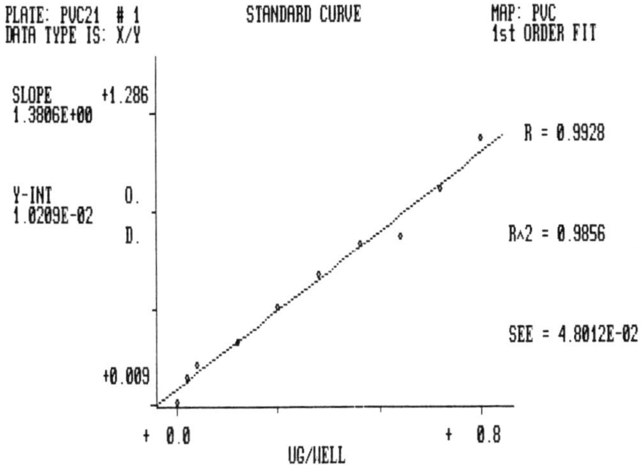

Figure 3. Typical standard curve for total CryIA ELISA. Density gradient purified CryIA(c) toxin protein crystals were solubilized, quantitated by chemical protein assay, and inputted to ELISA. Data were fitted by analysis with computer software (BioTek Instruments, Inc., Winooski, VT USA).

Table 1. Quantitative Assays for CryIA Toxin Protein Typical Data (Sample Values) for a BT Technical Powder

%(w/w) CryIA by ELISA		%(w/w) CryIA by PGel Assay	
mean value (N = 12)	standard deviation	mean value (N = 3)	standard deviation
23.13	1.42	18.00	0.20
20.87	0.04	20.60	1.20
22.01	0.89	17.13	0.25
20.79	0.05	21.77	0.68
24.34	0.22	20.60	1.10
19.96	0.04	17.15	0.24
21.85	0.56	21.85	0.51
21.84	1.03	22.40	1.40
22.90	0.38	15.33	1.38
21.82	0.64	19.72	1.21
23.27	0.07	21.86	2.57
19.96	0.15	18.11	2.60
24.38	1.81	24.53	0.03
18.62	2.70	22.25	0.78
20.05	0.75	17.90	0.33
26.20	1.05	17.17	0.99
21.37	1.47	18.51	0.50
23.50	1.00	20.40	1.00
16.67	1.15	17.59	1.02
22.67	0.18	18.58	1.02

ELISA:
grand mean (N = 20) = 21.81% CryIA
standard deviation (between-assays) = 2.172% CryIA
average within-assay standard deviation = 0.78% CryIA

PGel Assay:
grand mean (N = 20) = 19.57% CryIA
standard deviation (between-assays) = 2.37% CryIA
average within-assay standard deviation = 0.95% CryIA

Summary and Conclusions

The ELISA format described here works well for determination of total CryIA toxin protein in many types of BT production samples. The site specificity of our total CryIA primary antibody reagent overcomes problems caused by the presence of proteases and multiple, related CryIA toxin proteins in samples. This ELISA format, with nonspecific binding of solubilized CryIA toxin protein to the solid phase, is possible because levels of analyte are high in these samples. The nonspecific nature of this initial binding step presents potential problems, however, since heterologous proteins can compete for binding and some surface active compounds (*eg* detergents, surfactants) in samples can interfere with binding of CryIA toxin protein to the plastic. We are also investigating alternative ELISA formats that overcome these problems.

Acknowledgments

We thank Susan Brussock, Catrina Banks, and Yvonne Gattelli for the PGel Assay data in Table 1.

Literature Cited

1. Hofte, H.; Whiteley, H. R. Microbiol. Reviews 1989, 53 (2), 242-55.
2. Registration Standard for the Reregistration of Pesticide Products Containing *Bacillus thuringiensis* as the Active Ingredient, Environmental Protection Agency, Office of Pesticide Programs, Washington D.C., Case Number 0247.
3. Cheung, P. Y. K.; Hammock, B. D. in Biotechnology for Crop Protection; Hedin, P. A.; Menn, J. J.; Hollingworth, R. M. eds. 1988, ACS Symposium Series No. 379, pp. 359-72.
4. Wie, S. I.; Andrews, R. E., Jr.; Hammock, B. D.; Faust, R. M.; Bulla, L. A., Jr. Appl. Environ. Microbiol. 1982, 43, 891-4.
5. Smith, R. A.; Ulrich, J. T. Appl. Environ. Microbiol. 1983, 45, 586-90.
6. Wie, S. I.; Hammock, B. D.; Gill, S. S.; Grate, E.; Andrews, R. E., Jr.; Faust, R. M.; Bulla, L. A., Jr; Schaefer, C. H. J. Appl. Bacteriol. 1984, 57, 447-57.
7. Cheung, P. Y. K.; Hammock, B. D. Appl. Environ. Microbiol. 1985, 50, 984-8.
8. Faust, R. M.; Bulla, L. A., Jr. in Microbial and Viral Pesticides; Kurstak, E., ed.; Marcel Dekker, Inc., 1982, pp. 75-208.
9. Hofte, H.; Van Rie, J.; Jansens, S.; Van Houtven, A.; Vanderbruggen, H.; Vaeck, M. Appl. Environ. Microbiol. 1988, 54, 2010-7.
10. Towbin, H.; Staehelin, T.; Gordon, J. Proc. Nat. Acad. Sci. U. S. A. 1979, 76, 4350-4.

RECEIVED February 19, 1990

Chapter 11

The Light-Scattering Characterization of δ-Endotoxin Production in Inclusion Bodies

Fritz S. Allen[1,2], Betty J. M. Hannoun[3], Tammy B. Hebner[1], and Kathryn Nette[3]

[1]Acrogen Southwest Corporation, Albuquerque, NM 87109
[2]Department of Chemistry, University of New Mexico, Albuquerque, NM, 87131
[3]Mycogen Corporation, 5451 Oberlin Drive, San Diego, CA 92121

We describe the use of an instrument based on Multi-Parameter Light Scattering (MLS) technology which has the capability to measure the manner in which different optical polarizations are interchanged when polarized light is scattered by an object. This information is presented in a light scattering profile which is complex and has great capacity to characterize or fingerprint the scatterer. We report the application of this system to the monitoring of the growth of inclusion bodies in *Pseudomonas fluorescens* transformed with genetically engineered plasmids. These cells produced inclusion bodies composed largely of δ- endotoxin from *Bacillus thuringiensis*. The light scattering profiles of these cells containing inclusion bodies were compared with the host cells without the plasmid and resulting inclusion bodies. Using MLS technology, it was possible to differentiate between the cells. The instrument requires only a few minutes for collection and analysis of data. The optimization of expression levels by conventional means is time-consuming and tedious. Application of MLS technology to fermentation process development and control offers the capability to directly monitor the formation of inclusion bodies.

Optimizations of fermentations for recombinant protein production require rapid, reliable assays for recombinant protein measurements. Current assays for δ- endotoxin production in recombinant cell inclusion bodies require lengthy extractions and analyses. An "instantaneous" measurement of δ- endotoxin production could be used to optimize fermentation parameters. Inclusion body quantitation results during a batch fermentation would allow control of dissolved oxygen, pH, and nutrient feeds over time for maximal protein accumulation. The harvest time of a fermentation could also be optimized separately for each batch. Multi-Parameter Light Scattering is a rapid method of measuring inclusion body formation and growth and may be employed in the future for fermentation optimization.

Multi-Parameter Light Scattering

Multi-Parameter Light Scattering (MLS) technology describes the interaction of polarized light with a particle suspension (1). The polarization state of light can be described by four parameters. These parameters are: the total intensity of the light beam, the fraction of that intensity which is plane polarized, the orientation axis of the plane polarized component, and the fraction of the total intensity which is circularly polarized. When light passes through a

0097–6156/90/0432–0098$06.00/0

medium, the medium can alter any and all of the four polarization parameters. Thus, the polarization state of the beam which emerges from the medium may differ from that of the incident beam. The medium can cause any of the incident polarization parameters to contribute to any of the emergent polarization parameters (2). In general, there are sixteen coefficients which describe the effect of each of the incident polarization parameters on every one of the emergent polarization parameters. These sixteen coefficients constitute the Mueller matrix of the medium (3). Each medium has a different Mueller matrix and, furthermore, this matrix changes with the physical arrangement of the medium as well as the angle between the incident and emergent beams. Thus, while the matrix is complicated, it is very valuable because it contains all possible linear optical information about the medium.

The block diagram of the Mueller matrix instrumentation can be seen in Figure 1. The light from the laser is linearly polarized and this polarization state is altered by the polarization modulator. The period of the modulator is 20 microseconds. Thus, a complex sequence of polarizations is repeated in the incident light to the sample every 20 microseconds. On the movable arm which gathers data at a variety of scattering angles, there is a second polarizer and modulator. The two optical components act as a polarization sensitive gate to the detector. The period of the second modulator is 19 microseconds. During this time all different polarization states of the emergent light are gated to the detector. Thus, during 20 times 19 microseconds (380 microseconds), all combinations of incident and emergent light have been examined. This gives the system the capability to measure the way in which the polarization state of the light is altered during the scattering process. The different polarization exchange processes measure the Mueller matrix of the scatterer.

The optical signal incident on the photomultiplier detector is complex in the time domain. By exploring the specific time dependence of each of the terms in the final intensity expression given above in the theory, one can show that each matrix element will occur in this complex signal at a characteristic frequency. Fourier transformation or alternatively various electronic detection methodologies can be employed to isolate and determine the matrix elements.

The entire instrument is computer driven. A personal computer gathers the data and steps the motor arm to the next scattering angle of interest. At each scattering angle there is a waiting period for the arm to come to a complete stop. Then at least three measurements are taken, averaged and normalized for each element of the Mueller matrix. The data is stored on floppy disks and the motor arm moves to the next scattering angle. A scan from 120° to 16° in 4-degree increments takes approximately 6 minutes. The instrument uses four of the sixteen elements for the characterization of a sample. Each of the four elements is measured at a variety of scattering angles, with each such element-angle combination constituting a "test." The value determined for a sample on each of the tests constitutes the fingerprint, or signature, of the sample. The signature is determined by the size and shape of the organism and by the distribution and type of polarizable components within the organism. The MLS technique is sensitive to the internal physical arrangement of the macromolecular constituents of the cell. The size, the shape and the internal order are the physical basis of the signature that defines specifically the state of the organism under investigation.

Materials & Methods

MLS Instrument. The Signature One, (Acrogen Southwest, Corp., 3700 Osuna, NE, Albuquerque, NM 87109), was used in the experiment. The fully computerized instrument moves a detection arm upon which are mounted optics for measuring the various scattering signals through an angular range from 120° to 16° as measured from the emergent beam. The scattered intensity for the four matrix elements (1,1), (1,2), (3,4), (4,4) are measured with a 1-second integration time. These signals are measured three times, and the average signal and its deviation is determined. If the deviation of any element is greater than 10 %, a second set of three data points is obtained and averaged with the first three points. When the data has been obtained at 120°, the arm is automatically moved to 116° and data collection continues as before. Data is collected in 4°-steps from 120° to 16°. The angular aperture for scattered light entering the collection optics is set at 1.5°. The data set files can be plotted to provide hard copy display for each matrix element. This instrument was previously described and will only briefly be discussed here (4). The instrument uses a disposable sample container to facilitate sample introduction into the optical compartment. The bacterial suspension was filtered through a 10-micron membrane filter into the optical cuvette. The volume of sample in the cuvette was 1.5 ml. The filtration reduces the clumps of bacteria or debris which can enter the optical compartment and cause spurious signals. The suspension of bacteria was placed in the disposable sample container, which was then placed in the instrument.

P. fluorescens Cells. *Pseudomonas fluorescens* cells transformed with a plasmid for the production of *Bacillus thuringiensis* δ- endotoxin were used in these experiments along with wild type parent cells. Wild type *P. fluorescens* and transformed *P. fluorescens* cells were grown in peptone-salts medium in shake flasks. Recombinant cells were induced for δ-endotoxin production after 24 hours of growth. Four wild type and four recombinant fermentations were conducted at the same temperatures and aeration conditions. Throughout each fermentation, samples were aseptically removed at 24, 36, 48, 55.7, 72.2, and 79.2 hours. After plating on L agar for viable cell counts, samples were fixed with Lugol's iodine. Fixation with Lugol's iodine killed all viable cells, while maintaining cell integrity.

Samples were diluted 1:100 in PBS (pH=7.2) and vortexed gently for approximately 10 seconds. The samples were centrifuged at 3000 rpm for 10 minutes. Following centrifugation, the pellet was resuspended with 10 ml PBS and recentrifuged at 3000 rpm for 10 minutes. The pellet from the second centrifugation was resuspended in 10 ml PBS. The samples were further diluted 1:10 with PBS and filtered through a 10 micron filter and a reading of 2.20 was obtained on an Autobac Multitest System Photometer (model 106M, Pfizer Diagnostics Division, Pfizer Inc., New York, NY). Wet mount preparations were prepared and the samples were checked for inclusion bodies using a Leitz Laborlux D (Wild Leitz USA, Inc.) microscope with a 40X objective. Each sample was analyzed in triplicate.

Data Presentation and Generation of Plots. Four different elements of the scattering matrix were plotted as a function of scattering angle: elements (3,4), (1,2), (4,4) and voltage or (1,1). In order to assess the changes in the optical signatures as the inclusion bodies are formed, we have defined the integrated Mueller matrix difference parameter. This parameter was obtained by vector subtraction of the optical signature measured at the time of induction from those signatures which were determined subsequently. The sum of the squares of the elements of the residual signature gives the parameter. The parameter gave a convenient means to measure the changes in the optical signature with a single value. When this value was plotted against time of incubation, a production curve for the inclusion bodies was obtained.

Results & Discussion

Wild type *P. fluorescens* and plasmid-transformed *P. fluorescens* fermentations were characterized using Multi-Parameter Light Scattering technology. Four matrix elements, (3, 4), (1, 2), (4, 4), (volt), were determined as a function of scattering angle. The data indicate that at the time of induction, 24 hours, there are already some differences between the recombinant and wild type. The differences are illustrated with the (3, 4) matrix element in Figure 2. The wild type and recombinant variations may be due to pre-induction expression of recombinant protein. A protein band too faint to quantify was visible in this sample using polyacrylamide gel electrophoresis.

Following induction of plasmid-containing cells, the signatures of the recombinant cells varied significantly with time. Some variations in the wild type signatures also occurred, but the small variations in wild type signatures can be attributed to normal changes in cell composition during stationary phase, such as depletion of anabolic precursors and utilization of carbohydrate and phosphate reserves. Similar changes would also be occurring in recombinant cells during stationary phase, contributing to recombinant signature variations over time. Since the signature variations with recombinant cells are much larger than the wild type signature variations, additional physiological changes, such as inclusion body formation, are being detected in the recombinant cells.

Four fermentations with wild type cells and four fermentations with plasmid-containing cells were conducted to determine the reproducibility of the signatures over replicate experiments. The signatures of three out of four wild type fermentations were virtually the same at every time point. The fourth fermentation had somewhat different signatures. Figure 3 shows the (3, 4) matrix elements at 55.7 hours for all four fermentations with wild type cells, illustrating deviations in one fermentation experiment. Viable cell counts taken

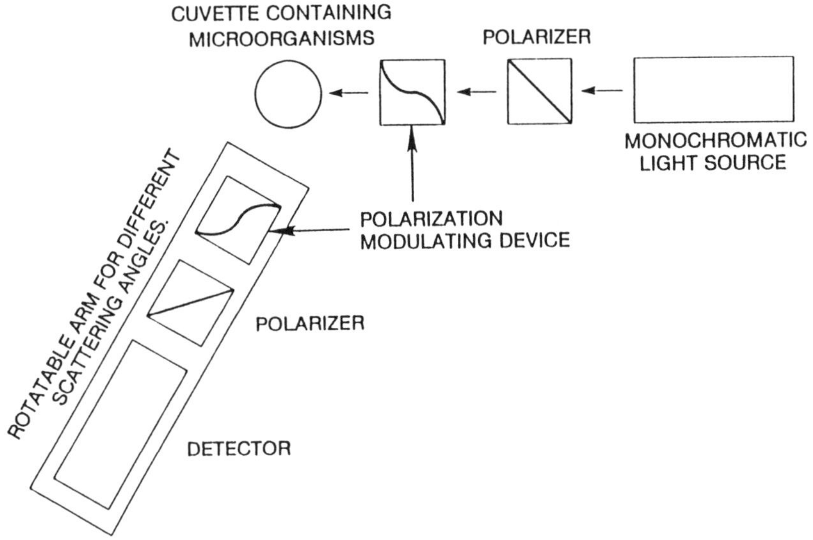

Figure 1. Schematic diagram of Multi-Parameter Light Scattering instrument.

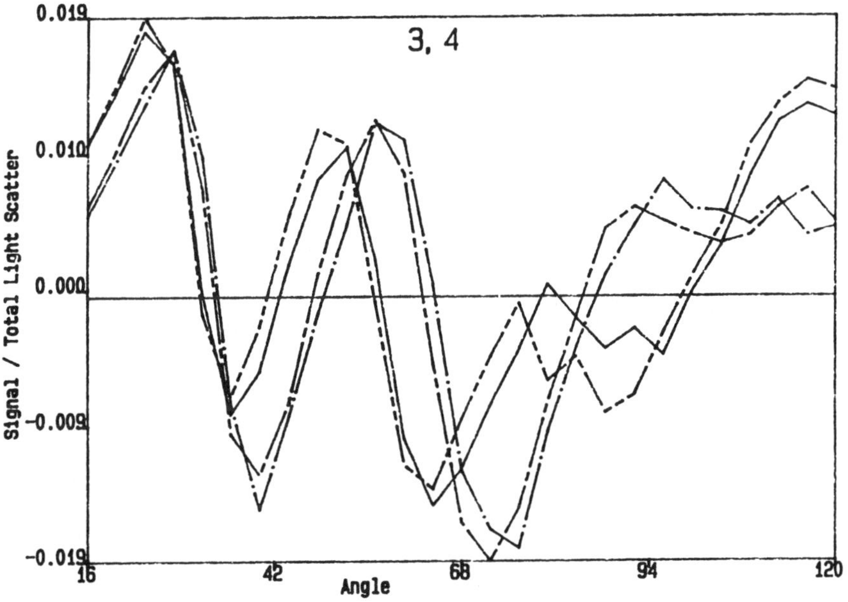

Figure 2. (3,4) Mueller matrix element of *P. fluorescens* as a function of scattering angle for two fermentations with wild type cells and two fermentations with plasmid-containing cells at 24 hours: (———) wild type run 1, (—— --) wild type run 2, (———— -) plasmid-containing run 5, (———— ·) plasmid-containing run 6.

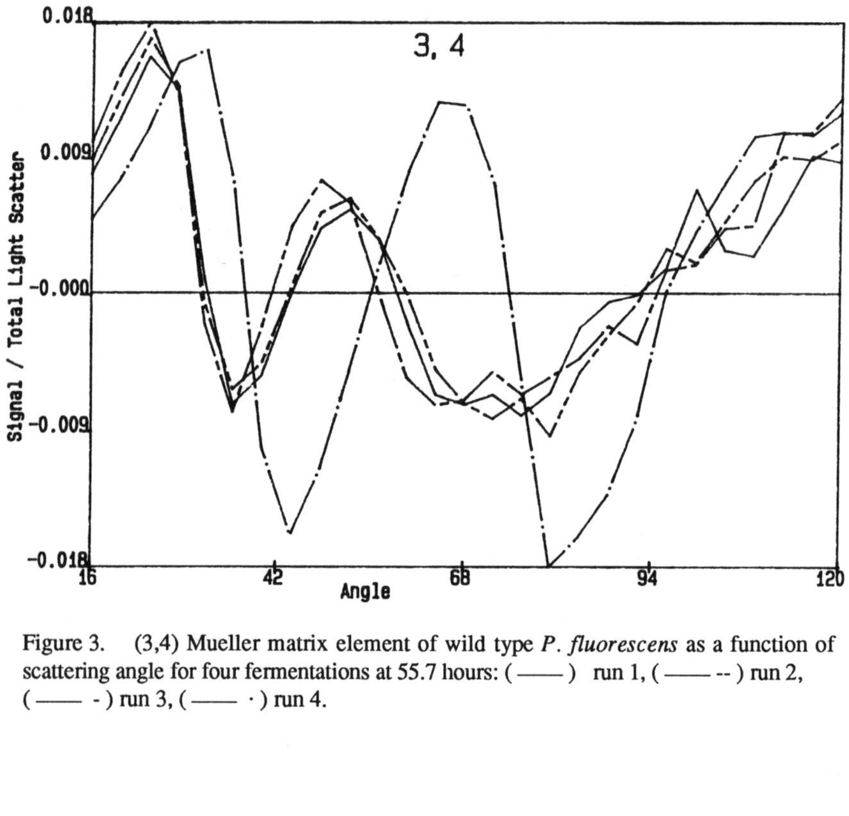

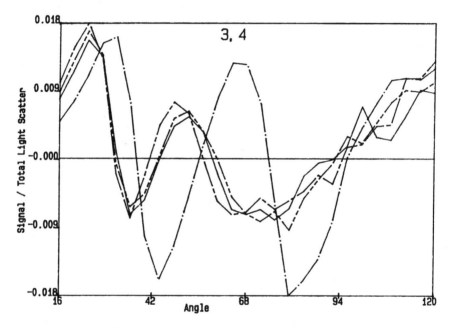

Figure 3. (3,4) Mueller matrix element of wild type *P. fluorescens* as a function of scattering angle for four fermentations at 55.7 hours: (——) run 1, (—— --) run 2, (—— -) run 3, (—— ·) run 4.

during the four wild type and four recombinant fermentations are shown in Table I. The three wild type fermentations that had similar signatures also had nearly constant viable cell counts between 24 and 79.2 hours. The viable cell counts for the fermentation with a different signature decreased by more than a factor of 10 between 55.7 and 79.2 hours. That particular fermentation appears to have completed stationary phase and entered the death phase earlier than the other three fermentations. Similar results occurred with the four recombinant fermentations. Three of the four recombinant fermentations gave similar signatures at all time points. The fourth recombinant fermentation had a significantly different signature. The three recombinant fermentations with similar signatures had nearly constant viable cell counts between 24 and 79.2 hours. The fourth recombinant fermentation with a different signature showed a 100-fold decrease in viable cell counts between 55.7 and 79.2 hours. Furthermore, as shown in Table II, the recombinant protein concentration peaked at 55.7 hours in the fermentation with a different signature instead of at 72.2 hours as in the three recombinant fermentations with similar signatures. These observations indicate that reproducible fermentations lead to reproducible signatures. Furthermore, fermentations during which cells enter death phase early and/or toxin production peaks early give significantly different signatures than normal fermentations.

Table I. Normalized Cell Counts Over Time

Sample Number	1	2	3	4	5	6	7	8
Time (hr)		Wild Type				Plasmid-Containing		
24.0	2.1	2.9	2.1	5.1	2.9	2.4	2.4	5.9
36.0	2.0	4.3	2.3	3.7	2.9	2.4	2.8	3.8
48.0	2.5	-	2.0	4.8	2.7	2.9	2.4	3.9
55.7	1.7	2.1	2.3	4.8	2.2	2.2	1.8	2.4
72.2	2.3	1.5	2.1	0.95	1.2	1.5	1.0	0.20
79.2	2.0	2.0	2.2	0.17	1.8	2.0	1.5	0.03

Table II. Normalized δ- Endotoxin Concentrations per Cell

Sample Number	5	6	7	8
Time (hr)		Plasmid-Containing		
24.0	-	-	-	-
36.0	0.06	0.04	0.07	0.06
48.0	0.12	0.08	0.15	0.13
55.7	0.27	0.22	0.30	0.23
72.2	0.37	0.38	0.41	0.16
79.2	0.33	0.34	0.34	0.14

Production curves were constructed for each fermentation from the optical signatures at various time points. The production curve is a plot of the integrated Mueller matrix difference parameter over time and is thus a measure of the change in the samples over time. The production curves for the four wild type and four recombinant fermentations are shown in Figure 4. There are only small changes in the production curves for the wild type fermentations over time. The recombinant production curves show greater changes over time. The fastest rate of recombinant production curve change was between 24 and 36 hours, when the inclusion bodies were beginning to form. From 36 to 72.2 hours, the recombinant production curves increased approximately linearly with time. Three of the four recombinant fermentations showed a decrease in the production curve from 72.2 to 79.2 hours, whereas all four recombinant samples showed a decrease in δ- endotoxin concentration.

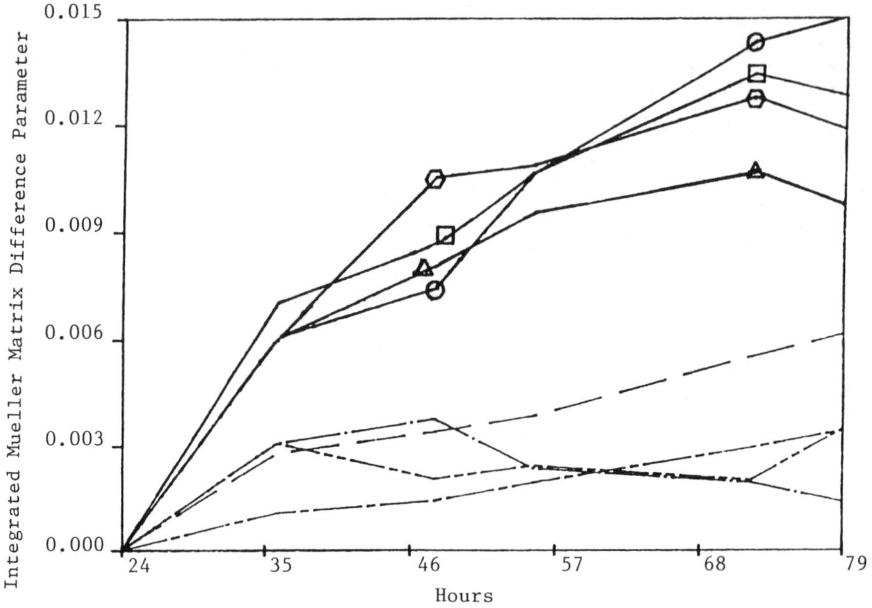

Figure 4. Integrated Mueller matrix difference parameter as a function of time for four fermentations with wild type cells and four fermentations with plasmid-containing cells: (——— --) wild type run 1, (——— –) wild type run 2 (——— ·) wild type, run 3, (—— ——) wild type run 4 (□) plasmid-containing, run 5 (○) plasmid-containing run 6, (⬡) plasmid-containing run 7, (△) plasmid-containing run 8.

Conclusion

Multi-Parameter Light Scattering technology was used to quantitate inclusion body formation and growth in recombinant cells. Significant changes in the optical signatures of recombinant cells occurred over time as the inclusion bodies formed and increased in size. Much smaller changes in the optical signatures of wild type cells were found over the same period of time. MLS technology can be used to rapidly characterize recombinant protein production in bacterial cell inclusion bodies and may be useful for fermentation optimization.

Literature Cited

1. Bottiger, J. R.; Fry, E. S.; Thompson, R. C. Appl. Optics 1980, 19, 1323-32.
2. D. R. Huffman; Hunt, A. J. Appl. Phys. 1974, 14 , 435-40.
3. Mueller, H. J. Opt. Soc. Am. 1948, 38, 661.
4. Newman, C.D. Ph.D. Thesis, University of New Mexico , New Mexico, 1987.

RECEIVED February 1, 1990

Chapter 12

Quantification of *Bacillus thuringiensis* Insect Control Protein as Expressed in Transgenic Plants

Roy L. Fuchs, Susan C. MacIntosh, Duff A. Dean, John T. Greenplate, Frederick J. Perlak, Jay C. Pershing, Pamela G. Marrone, and David A. Fischhoff

Plant Science Technology, Monsanto Agricultural Company, 700 Chesterfield Village Parkway, St. Louis, MO 63198

Quantitative assays were developed to analyze gene products produced by transgenic plants that express *Bacillus thuringiensis* insect control proteins. Initially *B. thuringiensis* proteins were produced in plants at approximately 10,000 fold lower specific activities than the same proteins produced by microbial fermentations. Consequently, accurate and selective assays were developed to quantitate *B. thuringiensis* specific DNA (gene copy), mRNA and protein produced by insect tolerant transgenic plants. The 100 fold increase in *B. thuringiensis* protein expression achieved in plants, based on these analysis, enabled the production of transgenic plants that are protected against agronomically important insect pests.

Bacillus thuringiensis (*B.t.*) strains represent the major microbial biocontrol agents used as agricultural insecticides (1). The potency, specificity and environmental safety of these organisms and their insecticidally active crystal proteins make *B.t.* an attractive insecticidal candidate for agricultural use. Unfortunately, the manufacturing costs, the short half life under field conditions and the slow kill rate of microbial formulations have hampered the widespread application of these biocontrol agents. Engineering the gene(s) encoding the insecticidal *B.t.* proteins directly into plants provides an agronomically and economically attractive alternative to microbial delivery systems. Genes encoding the *B.t.* proteins have been genetically engineered and expressed in tobacco (2,3,4) and tomato (5). Extremely low levels of the lepidopteran-active *B.t.* var. *kurstaki* (*B.t.k.*) proteins (<0.001% of total soluble plant protein) were produced in these plants (3) which control the most sensitive lepidopteran insects such as *Manduca sexta* (tobacco hornworm). Significant increases in *B.t.k.* expression were required for commercial level control of the less sensitive, agronomically important insects such as *Heliothis zea* (cotton bollworm). Recent improvement in *B.t.k.* gene expression has resulted in increased production of the *B.t.k.* protein in plants to up to 0.1%, a level which is sufficient to control even the least sensitive of the lepidopteran insects.

Biochemical characterization of transgenic plants that express the *B.t.k.* protein required development of new analytical assays to detect and quantitate the levels of *B.t.k.* proteins in plants. Analytical methods such as SDS-PAGE (6) and direct ELISA (7) assays were used to characterize microbially produced *B.t.k.* proteins since 30 to 50% of the total protein was *B.t.k.* protein. Development of assays for transgenic plants where the *B.t.k.* protein comprised only 0.001% of the total protein required much more sensitive and selective assays. Production of insect resistant transgenic plants depends not only on the insertion of the gene encoding the *B.t.k.* protein, but also transcription and translation into protein.

0097–6156/90/0432–0105$06.00/0

Analytical assays were needed to evaluate each of these events to optimize insecticidal efficacy.

This manuscript details many of the techniques that we have used to detect and quantitate *B.t.k.* expression in transgenic plants. Assays are described which establish: (a) the presence of the *B.t.k.* gene in plants, (b) the level of *B.t.k.* specific mRNA, (c) the level of *B.t.k.* protein and (d) insecticidal efficacy. All four types of data have been critical to overcome the extremely low, initial levels of *B.t.k.* expression in plants and to achieve dramatic increases in the insecticidal efficacy of these plants. The specific assays described should also serve as appropriate analytical tools to support the regulatory approval of insect tolerant transgenic plants for commercial use.

Presence of the *B.t.k.* Gene in Transgenic Plants

One or more copies of the gene(s) encoding *B.t.k.* protein, accompanied by the kanamycin antibiotic resistance marker gene, has been randomly inserted into the chromosome of the target plant via *Agrobacterium*-mediated DNA transfer to produce insect resistant transgenic crops. The specific site of insertion into the plant chromosome is a factor which can dictate the level of protein expression obtained from any insertion event. This phenomenon is known as "position effect" (8). Jones et al. (8) reported up to a 200 fold range in mRNA levels from the same gene depending upon the specific site of insertion. There is no direct correlation between the number of gene copies inserted and the level of *B.t.k.* protein expression. However, knowledge of the number of independent genes inserted is important to determine gene stability and to enable plant breeders to efficiently move the *B.t.k.* gene into other crop varieties.

Selection on antibiotic media (9), analysis for the antibiotic marker gene using functional assays (10) or detecting the *B.t.k.* gene using the polymerase chain reaction (PCR) technique (11) have been used to establish the presence of the *B.t.k* gene. However, these techniques do not provide quantitative data on the number of gene copies inserted. Southern blot analysis (12) provides quantitative data, particularly the ability to identify plants that contain a single versus mulitple insertions. Single insertion events are the most frequent and preferred event for plant breeding purposes.

We analyzed a number of transgenic tomato plants prior to the first field test in 1987 by Southern analysis to determine the number of independent *B.t.k.* gene insertions. Total DNA from tomato plants was digested with the restriction enzyme HindIII, which cleaves twice within the *B.t.k.* structural gene. The internal fragment of the *B.t.k.* gene is shown by the arrow in Figure 1. Two fragments comprised of a portion of the *B.t.k.* gene fused to the adjacent plant chromosomal sequences was also generated. Cleavage of each independent insertion produces two fragments of unique sizes, dependent on the nearest HindIII site in the adjacent plant DNA. The number of insertions was estimated by counting these fusion fragments. A Southern hybridization analysis in which the total HindIII digested tomato DNA was size separated using agarose gels, blotted to a membrane and hybridized to vector DNA is shown in Figure 1. Examples of plants with single insertion events, which show two unique fusion fragments, are shown in lanes 3 and 4. In lanes 1 and 2 are examples of plants with two independent insertions, indicated by the four fusion fragments. The non-transformed control showed no hybridizing bands (data not shown). Plants containing the single insertion showed the expected Mendelian segregation pattern and were selected for subsequent genetic crosses.

Levels of *B.t.k.* mRNA in Plants

The level of *B.t.k.* protein produced in transgenic plants depends on both the level of *B.t.k.* mRNA and translational efficiences of this mRNA. Assays to quantitate the amount of *B.t.k.*-specific mRNA were required to understand the molecular basis for the extremely low levels of *B.t.k.* mRNA and protein expression observed with the initial transgenic plants. Northern (5) and S1 nuclease protection (13) assays have been routinely used to quantitate the levels of mRNA. Northern assays can be performed using either total RNA or polyA-selected RNA. The latter contains primarily mRNA, which comprises approximately 2 to 4% of the total RNA, and therefore increases the assay sensitivity for *B.t.k.* specific mRNA by approximately 50 fold. This increased sensitivity was confirmed by comparing northern assays run with 40 μg of total RNA or 20 μg of polyA-selected RNA from tissue from several plants (Figure 2). Northern assays provide both a quantitative estimate of the amount of *B.t.k.*-specific RNA and an estimate of the proportion of the mRNA that is full length, capable of producing biologically active *B.t.k.* protein. Note that although the vast majority of mRNA detected in lanes 6, 7 and 8 represents full length, functional gene products, there are several minor, truncated RNA products that probably result from either premature termination, incorrect processing or nuclease cleavage. Once the majority of mRNA has been established to be full length, an S1 nuclease protection assay provides a rapid, easier assay that requires less tissue for comparison of mRNA levels than northern assays. A 300 base pair oligonucleotide fragment homologous to nucleotides 1 through 71 of the *B.t.k.* structural gene and containing a 229 base pair non-homologous tail was used as the hybridization probe for analysis by the S1 nuclease protection assay. The quantity of *B.t.k.* specific mRNA was estimated from densitometric scanning of the labelled protected fragment after separation on a sequencing gel. The S1 nuclease protection assay is comparable in sensitivity to the polyA-selected northern (Figure 3).

Levels of *B.t.k.* Protein

B.t.k. protein levels establish whether changes made at the molecular level have altered the level of insecticidal efficacy. Insecticidal efficacy, in turn, determines the agronomic potential of transgenic plants engineered for insect resistance. Unfortunately, analysis of the insecticidal efficacy of whole transgenic plants or plant parts are limited relative to microbial formulations where multiple dilutions can be made to determine specific and accurate LC_{50} values. Plant materials can be homogenized and incorporated into insect diets for direct feeding studies, but the extremely low levels of *B.t.k.* protein expressed in the initial transgenic plants precluded the use of this approach. We have used a detached leaf bioassay with a variety of lepidopteran insects that have very different sensitivities to the *B.t.k.* protein to provide a semi-quantitative estimation of the levels of *B.t.k.* protein in transgenic plants. Seven lepidopteran insects were tested for their relative sensitivities to purified *B.t.k.* (CryIA(b)) protein encoded by the same gene as expressed in the transgenic plants (MacIntosh, S.C. et al., J. Invert. Path., in press). Full length protein was purified from *Escherichia coli* cells containing this gene and the activated fragment produced and purified essentially as described by Beegle (14). LC_{50} values refers to the quantity of *B.t.k.* protein, in μg/ml, that was required to kill 50% of the neonate insect larvae in the diet incorporation assay described by Marrone et al. (15). A gradation of insect sensitivities was observed. The least sensitive insects, *Spodoptera exiqua* (beet armyworm) and *Heliothis zea* (cotton bollworm) were approximately 20 fold less sensitive to this *B.t.k.* protein than *H. virescens* (tobacco budworm). Tobacco budworm was 40 fold less sensitive than the most sensitive insect, *Manduca sexta* (tobacco hornworm). These four insects span 800 fold in sensitivity and enabled rapid screening of transgenic plants for semi-quantitative estimates of the level of *B.t.k.* protein produced *in planta.*

Transgenic tomato plants were produced which contain each of three different genes encoding *B.t.k.* protein. Plants derived from the vector pMON9921 contain a wild type truncated *B.t.k.* gene comparable to those described by Fischhoff et al. (5). Plants derived from vectors pMON5370 and pMON5377 contain *B.t.k.* genes that have modifications within the structural gene for *B.t.k.* that led to dramatic increases in the expression of the *B.t.k.* protein. Only the nucleotide sequence was changed, not the amino acid sequence of the *B.t.k.* proteins. Leaf tissue from transgenic tomato plants containing these three genes was assayed with tobacco hornworm (Figure 4) and beet armyworm (Figure 5) to ascertain the relative levels of *B.t.k.* protein. Leaves from all three transgenic plants were completely protected against tobacco hornworm damage, whereas the leaf from the non-transgenic control was consumed (Figure 4). The transgenic plant containing pMON9921 was only marginally protected against beet armyworm damage, with the plants containing pMON5370 and pMON5377 showing total protection (Figure 5). Based on the relative insect sensitivities to purified *B.t.k.* protein, the plants containing pMON5370 and pMON5377 would be predicted to express at least 100 fold higher quantities of *B.t.k.* protein than the plant containing pMON9921.

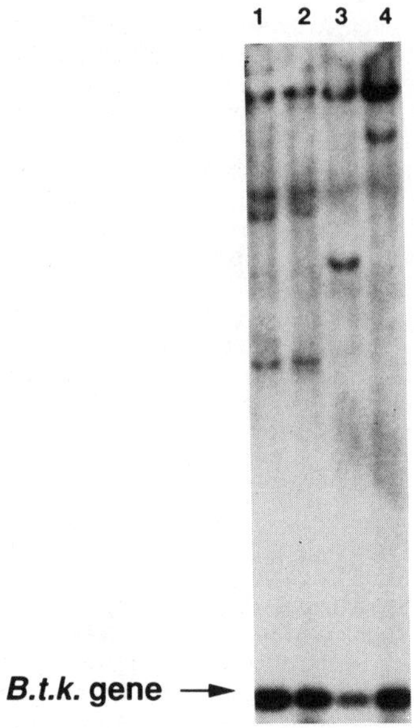

Figure 1. Southern hybridization. Autoradiograph of the nitrocellulose filter after hybridization of HindIII-cleaved total DNA from four independent transgenic plants to a *B.t.k.* gene probe. The arrow shows the internal HindIII fragment of the *B.t.k.* gene.

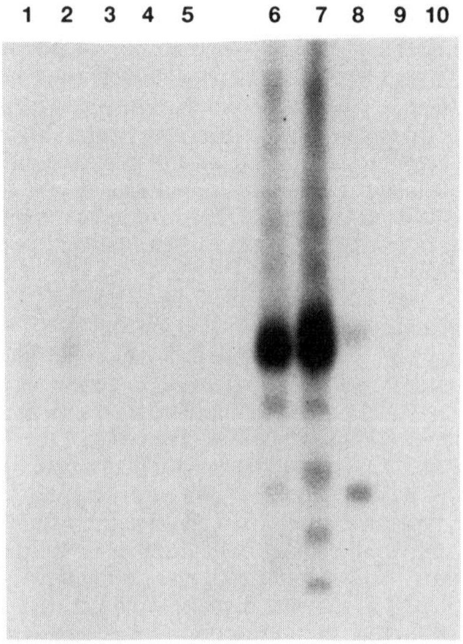

Figure 2. Northern hybridization. Autoradiograph of the nitrocellulose filter after hybridization of total (lanes 1 to 5) or polyA-selected (lanes 6 to 10) RNA from the same 5 independent transgenic tomato plants to a *B.t.k.* gene probe.

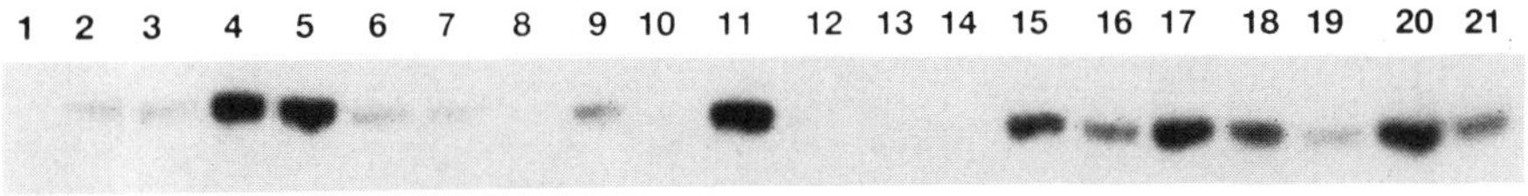

Figure 3. S1 nuclease protection. Autoradiograph of the RNA fragment protected from S1 nuclease digestion by a *B.t.k.* gene probe. Total RNA (40 μg) from 20 independent tomato plants containing the pMON9921 gene construction was analyzed by the S1 nuclease protection assay.

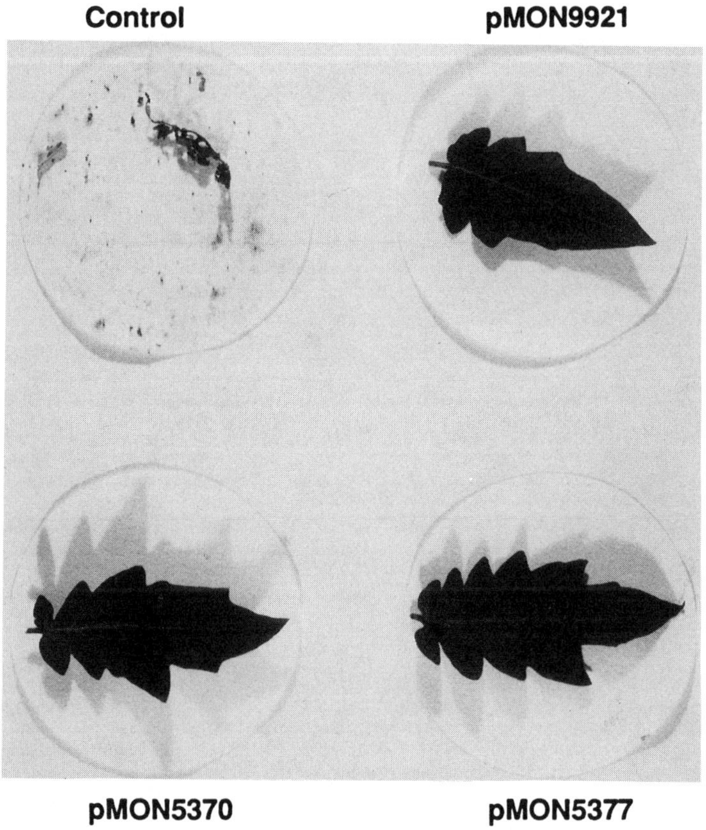

Figure 4. Tobacco hornworm bioassay. Detached leaves from control or transgenic tomato plants containing pMON9921, pMON5370 or pMON5377 were bioassayed against tobacco hornworm larvae as described (5).

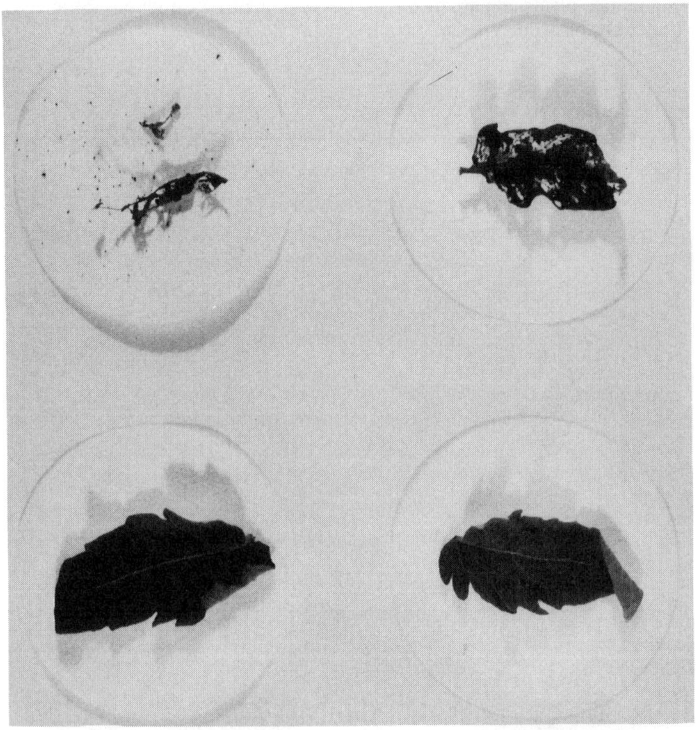

Figure 5. Beet armyworm bioassay. Detached leaves from control or transgenic tomato plants containing pMON9921, pMON5370 or pMON5377 were bioassayed against beet armyworm larvae as described (5).

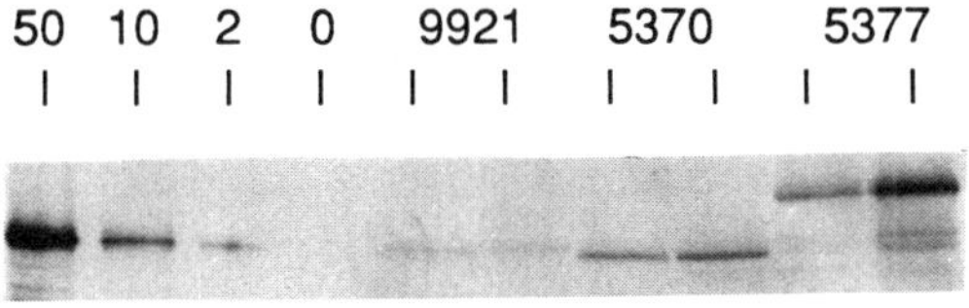

Figure 6. Western blot. The amount of *B.t.k.* protein in tomato plants was determined essentially as described (3). Extracts (50 μg of total protein) of a non-transgenic tomato control was spiked with 50, 10, 2 and 0 ng of purified *B.t.k.* protein. Extracts were prepared and analyzed from two independent transgenic tomato plants containing pMON9921, pMON5370 and pMON5377.

The level of *B.t.k.* protein in these plants was more precisely determined by homogenizing the transgenic plant tissue and combining the extract in the diet incorporation biossay. Data from these experiments confirmed the data from the detached leaf bioassays, that plants containing pMON5370 or pMON5377 expressed approximately 100 fold more *B.t.k.* protein than the plants containing pMON9921.

Two quantitative assays for the direct detection of *B.t.k.* proteins expressed in transgenic plants are the ELISA and western blot assays. Whereas a direct binding ELISA is typically used to quantitate the *B.t.k.* protein from microbial formulations (7), a double sandwich, indirect ELISA was required to quantitate *B.t.k.* protein from transgenic plant tissue (4). The extremely low levels of *B.t.k.* protein in transgenic plant tissue (0.001 to 0.1%) compared to *B.t.k.* formulations (30 to 50%) negated the efficient binding of the *B.t.k.* protein directly to the microtiter plates. Quantitative ELISA assays have been developed to estimate the levels of *B.t.k.* protein expression in plants. Although more labor intensive, the western blot analysis has been the analytical method of choice to date for quantitating *B.t.k.* protein levels in transgenic plants (3,4). *B.t.k.* protein quantities of 0.001% or 10 ng of *B.t.k.* protein per mg of total protein was detected in the early transgenic tobacco plants (3). As the expression levels have increased, the ability to detect and quantitate *B.t.k.* protein has become easier and more accurate. The level of *B.t.k.* protein in the tomato plants shown in Figures 4 and 5 was estimated by western analysis. Plants containing pMON9921 produced *B.t.k.* protein at approximately 0.001%, whereas the plants containing pMON5370 and pMON5377 produced *B.t.k.* protein at up to 0.1% of the total protein (Figure 6). The molecular weight difference in the *B.t.k.* proteins observed with the plants containing pMON5370 and pMON5377 was consistent with the size of the respective genes introduced.

Although the western analysis may be the most frequently used method to estimate *B.t.k.* protein levels in plants, it is the least sensitive of the protein assays described (Figure 7). Data in Figure 7 were based on an accumulation of assays generated on numerous plants containing the same vectors as the plants shown in Figures 4, 5 and 6. The tobacco hornworm bioassay on detached leaves was the most sensitive assay, followed by the ELISA and finally the western assay. One or more of these assays can be used to quantitate the level of *B.t.k.* protein expressed *in planta*, depending on the level of *B.t.k.* expression.

Plant Analysis

The repertoire of analytical assays described above provides the tools required to analyze transgenic plants. Multiple plants produced from any one gene construction or plants generated from different genes with changes in promotors, leader sequences, terminators or changes within the structural gene can be compared relative to transcriptional and/or translational efficiences. Figure 8 provides a schematic diagram of how we analyze transgenic plants derived from a new gene construction and how these plants are compared to previous plants containing different genes. The tobacco hornworm bioassay is used to identify the transgenic plants (usually 25 to 30%) that express the *B.t.k.* protein at levels sufficient to totally protect the leaf from damage. Western analysis is then used to identify the three or four highest expressing plants. These plants are compared to the highest

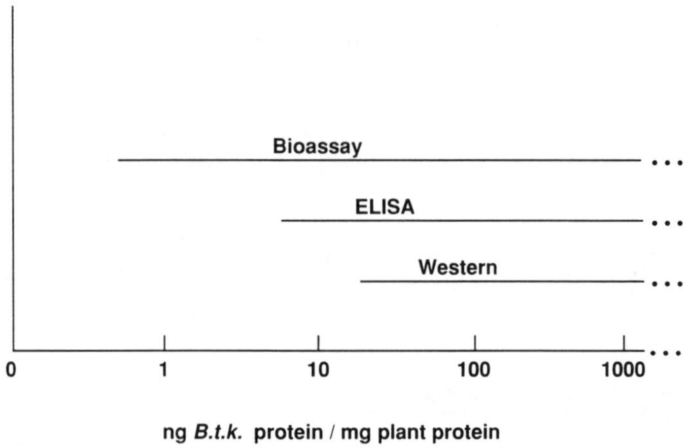

Figure 7. Protein assay sensitivities. The relative sensitivities of the three assays used to quantitate the amount of *B.t.k.* protein in transgenic tomato plants were compared.

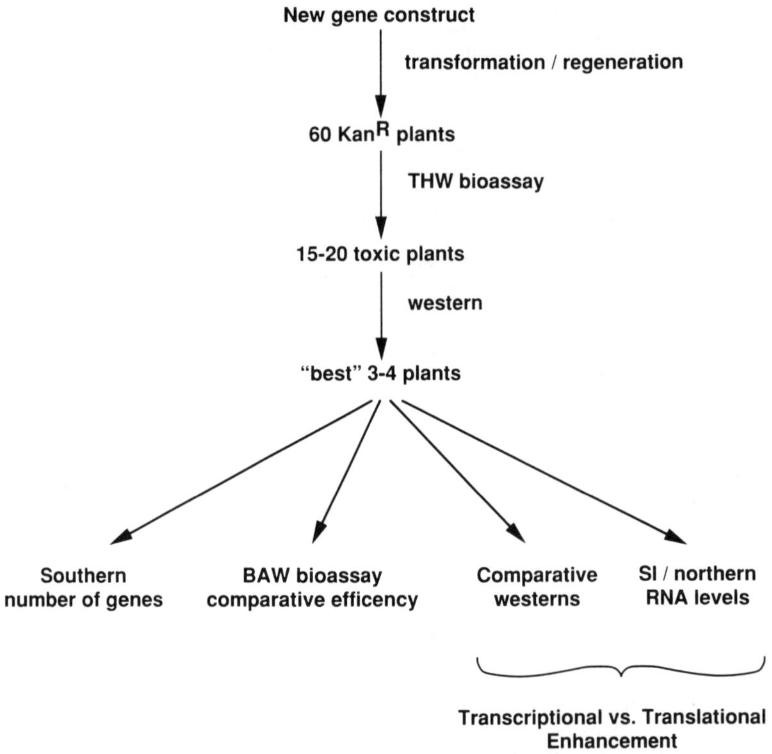

Figure 8. Plant analysis. Plants from a new gene construction are analyzed to identify the plants which express the *B.t.k.* protein at the highest levels. The protein levels in these plants are compared to the levels derived from other gene constructions to determine if changes made at the molecular level alter transcriptional and/or translational efficiencies.

expressing plants generated from other genetic constructions for: (a) the level of *B.t.k.* protein by western blot analysis or the beet armyworm bioassay and (b) mRNA levels by northern blot or S1 nuclease protection assays. Changes in *B.t.k.* RNA and/or protein levels establishes whether the genetic alterations engineered into new genes results in transcriptional and/or translational enhancements. Using this approach, we have engineered changes within the structural gene for *B.t.k.* (i.e. pMON5377) to enhance the *inplanta* expression of *B.t.k.* by over 100 fold compared to the wild type gene counterpart (pMON9921, Figure 7). This increase in expression has enhanced the insecticidal efficacy of tobacco, tomato and cotton transgenic plants to levels that provides effective control of the agronomically important lepidopteran insects.

Acknowledgment

The authors thank Jeanne Layton and Nancy Mathis for their transformation and regeneration support for tomato and tobacco, respectively and for the insect bioassay support from Evelyn White-McClain and Paul Gahr.

Literature Cited

1. Klausner, A. Bio/Technology 1984, 2, 408-419.
2. Adang, M.J.; Firoozabady, E.; Klein, J.; DeBoer, D.; Sekar, V.; Kemp, J.D.; Murray, E.; Rocheleau, T.A.; Rashka, K.; Staffield, G.; Stock, C.; Sutton, D.; Merl, D.J. In Molecular Strategies fro Crop Protection; C.J. Arntzen and C. Ryan, Eds.; Alan R. Liss, Inc., New York, New York. 1987; pp. 345-353.
3. Barton, K.A.; Whitely, H.R.; Yang, K.S. Plant Physiol. 1987, 85, 1103-1109.
4. Vaeck, M.; Reynaerts, A.; Höfte, H.; Jansens, S.; DeBeuckeleer, M.; Dean, C.; Zabeau, M.; Van Montagu, M.; Leemans, J. Nature 1987, 328, 33-37.
5. Fischhoff, D.A.; Bowdish, K.S.; Perlak, F.J.; Marrone, P.G.; McCormick, S.M.; Niedermeyer, J.G.; Dean, D.A.; Kusano-Kretzmer, K.; Meyer, E.J.; Rochester, D.E.; Rogers, S.G.; Fraley, R.T. Bio/Technology 1987, 5, 807-813.
6. Brussock, S.M.; Currier, T.C. ACS Symposium Series, this volume.
7. Groat, R.G.; Mattison, J.W.; French, E.J. ACS Symposium Series, this volume.
8. Jones, J.D.G.; Dunsmuir, P.; Bedbrook, J. EMBO J. 1985, 4, 2411-2418.
9. Fraley, R.T.; Rogers, S.G.; Horsch, R.B. CRC Crit. Rev. Plant Sci. 1986, 4, 1-46.
10. McDonnell, R.E.; Clark, R.D.; Smith, W.A.; Hinchee, M.A. Plant Mol. Biol. Reporter 1987, 5, 380-386.
11. Saike, R.K.; Scharf, D.; Faloona, F.; Mullis, K.B.; Horn, G.T.; Erlich, H.A.; Arnheim, N. Science 1985, 230, 1350-1354.
12. Southern, E. J. Mol. Biol. 1975, 98, 503-517.
13. Hanley-Bowdoin, L.; Elmer, J.S.; Rogers, S.G. Nucleic Acids Res. 1988, 16, 10511-10528.
14. Beegle, C.C.; Dalmage, H.T.; Wolfenburger, D.A.; Martinez, E. Environ. Entomol. 1981, 10, 400-401.
15. Marrone, P.G.; Ferri, F.D.; Mosley, T.R.; Meinke, L.J. J. Econ. Entomol. 1985, 78, 290-293.

RECEIVED February 19, 1990

Chapter 13

High-Performance Liquid Chromatography Analysis of Two β-Exotoxins Produced by Some *Bacillus thuringiensis* Strains

Barry L. Levinson[1]

Ecogen Inc., 2005 Cabot Boulevard West, Langhorne, PA 19047–1810

Reverse-phase HPLC separation has proven to be a valuable analytical tool for the detection and quantification of β-exotoxin production in *Bacillus thuringiensis* (BT) culture supernatants. The method is simple, rapid, sensitive and reliable in the detection of the classical β-exotoxin, described in the literature. Exotoxin production was assigned to a plasmid in six strains, from four varieties (1 *thuringiensis*, 3ab *kurstaki*, 9 *tolworthi*, and 10 *darmstadiensis*). The location of exotoxin genes on transmissible plasmids suggests that genetic material from such strains must be used only with circumspection, and that flagellar serotyping as a tool for determining the potential for exotoxin production is of limited value. A new exotoxin, Type II β-exotoxin, was discovered in BT strain HD-12 (serotype 8ab, variety *morrisoni*), and subsequently purified and partially characterized. This material is a more specific insecticide than Type I β-exotoxin, and is very active against the Colorado potato beetle, *Leptinotarsa decemlineata*. Identification of this new exotoxin demonstrates that bioassay needs to remain an integral part of the screening process for rigorous exclusion of exotoxin-producing strains from *Bacillus thuringiensis* products.

In addition to the proteinaceous δ-endotoxins, to which most of this volume is devoted, some strains of *Bacillus thuringiensis* (BT) produce a heat-stable insecticidal adenine-nucleotide analogue, known as β-exotoxin or thuringiensin (structure shown in Figure 1). Several excellent reviews cover the basic biology of β-exotoxin activity (1, 2) and chemistry (3, 4). The toxicity of this compound is thought to be due to inhibition of DNA-directed RNA polymerase by competition with ATP (5-7). Sublethal doses of β-exotoxin cause developmental abnormalities in insects, probably by the same mechanism. Because mammalian mRNA polymerases are quite sensitive to β-exotoxin (8, 9), the United States Environmental Protection Agency has been reluctant to sanction the

[1]Current address: Pharmacology Department, Berlex Laboratories, Inc., 110 East Hanover Avenue, Cedar Knolls, NJ 07927–2095

0097–6156/90/0432–0114$06.75/0

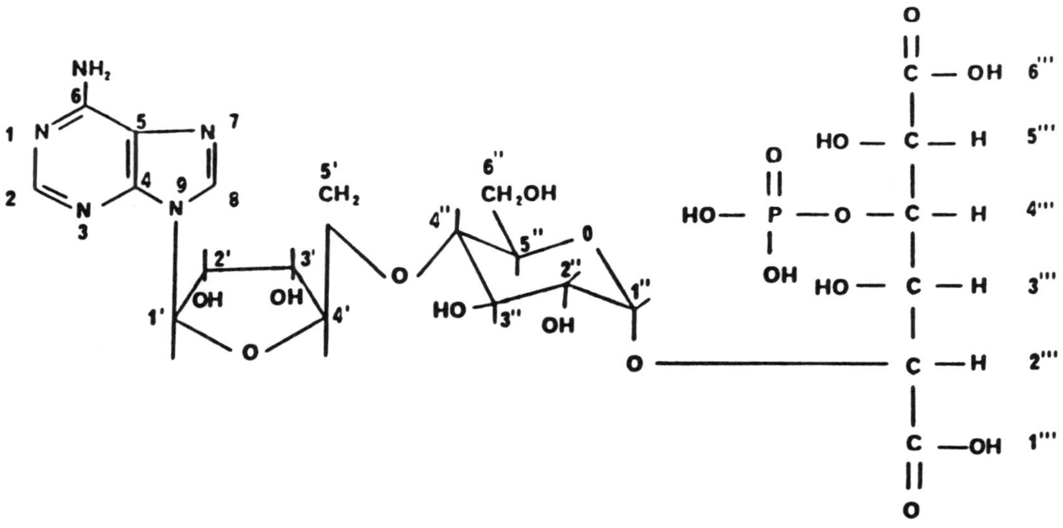

Figure 1. Structure of β-exotoxin. (Reproduced from Ref. 18. Copyright 1987 ACS.)

use of products containing this material (see Chapter 2 of this volume). Therefore, it is necessary to screen potentially useful new isolates of BT for β-exotoxin production.

Traditionally, testing for exotoxin has been done by bioassay of autoclaved culture supernatants, since exotoxin is stable to autoclaving at neutral pH. Because the common house fly, *Musca domestica* is very sensitive to exotoxin, but not at all to endotoxins, it has been the insect of choice for these assays (10, 11). This method, however, is time-consuming and not easily quantified. Purification of exotoxin standard has been done by means traditional for nucleotides, typically involving several steps--absorption onto activated carbon, precipitation of insoluble salts and ion exchange (12-14). This is also time-consuming.

In an effort to develop a rapid, sensitive method for the detection and quantification of β-exotoxin, several laboratories have used high performance-liquid chromatography (HPLC). The first such use was reported by Oehler, et al., and used a Waters C18 μBondapak reverse-phase column, with 0.1% trifluoroacetic acid in water as the mobile phase (15). However, other materials from crude culture filtrates coelute under these conditions (16). Bubenschikova, et al. used a Hitachi Type 2632 anion exchange resin with an acid and salt gradient elution at 50°C (17). This appeared to be specific, but was not very rapid, requiring 45 minutes for analysis, not including column recycling.

Campbell, et al. altered the Oehler method, using a pH 3 phosphate buffer instead of the TFA. They reported this to be both fast and specific for β-exotoxin, even in complex fermentation broths (18). We used a modification of their method for this study (Levinson, B.L., et al., J. Bacteriol., in press). Comparison of the results from HPLC with those of fly bioassay show that HPLC can be used to detect β-exotoxin production in different strains of BT, and to quantify its production, which varies dramatically between strains. This HPLC system can also be used in an essentially single-step purification procedure.

It may also be important to identify the location of genes involved in exotoxin production, so that they can be excluded from potential BT production strains. Analysis by HPLC was used to study genetically altered strains to identify the molecular location of the exotoxin gene(s). The data included in this study indicate that β-exotoxin production is plasmid-encoded in six BT strains, from varieties *thuringiensis* (serotype 1), *kurstaki* (serotype 3ab), *tolworthi* (serotype 9) and *darmstadiensis* (serotype 10). The plasmid responsible is a different size in each strain, and in five of the six cases, this plasmid also encodes at least one δ-endotoxin. A CryI-type (P1) crystal protein gene is on each plasmid, and sometimes also a CryII-type (P2) (19, 20).

For one strain, HD-12 (serotype 8ab, variety *morrisoni*), no β-exotoxin peak was found by HPLC, and yet fly toxicity was observed. Isolation and partial characterization of the toxic factor showed it to be a new exotoxin, which we termed β-exotoxin, Type II (Levinson, B.L., et al., J. Bacteriol., in press). This chapter reviews those data, and further describes the physical and biological characteristics of Type II β-exotoxin. As is true for Type I β-exotoxin, Type II exhibits a broad range of insecticidal activity; however, the insectidicidal specificity of Type II β-exotoxin is different from that of Type I.

Materials and Methods

Details on the strains studied, general HPLC methodology, and other methods can be found in Levinson, B.L., et al. (J. Bacteriol., in press). Additional methods not detailed there are given below. Our reisolated HD-12 is available from the Agricultural Research Culture Collection, Peoria, IL, as NRRL B-18354. Some derivatives of bacterial strains were generated at Ecogen: HD-537 strains were made by A. Macaluso; HD-13 and NRD-12 derivatives by J. González. NRD-12 was isolated from a 1985 Sandoz product, Javelin, Lot 51511. HPLC separations were performed on a Waters 840 system, using a Vydac

218TP54 column at 30°C, with a mobile phase of 50 mM KH_2PO_4, pH 3.0, at 2 ml/min, and detection at 260 nm. Fly bioassays were done by diet incorporation, as described (Levinson, B.L., et al., J. Bacteriol., in press); all other insect bioassays were by surface contamination, typically in cups yielding 6 cm^2 surface area.

HPLC purification of exotoxins. For preparative purposes, a mobile phase of 57 mM acetic acid, pH 3.0 (no adjustment) was found to give adequate separations on the Vydac 218TP54 wide pore C18 reverse phase column used for analytical purposes, although the retention times are somewhat earlier. This phase has the advantage over phosphate in product workup, since it can be removed completely by lyophilization. For a typical purification, a lyophilized powder of crude autoclaved culture supernatant was reconstituted at about 5x concentration in water. This material was passed through an Amicon (Danvers, MA) stirred cell, using a YM-2 (1000 MWCO) filter. The filtrate was collected and acidified to pH 3.0 with concentrated phosphoric acid. This material was repeatedly injected in 100 μl volumes, and fractions of about 0.6 ml were collected as pools. The fractions containing exotoxin (Type I or II) were identified by reinjection, and the desired fractions lyophilized. For Type I exotoxin, this results in a small contamination with dephosphorylated material, which apparently forms during the lyophilization step.

NMR and mass spectroscopy. ^{1}H-NMR was performed by Dr. Laurine Galya of Pittsburgh Applied Research Corporation on a 300 MHz Brucker spectrometer. Samples of purified exotoxins were resuspended in water and lyophilized to remove residual acetic acid, then taken up in d_6-DMSO (99.9 atom% d, from Sigma), containing 0.03% TMS, and acidified with a small amount of acetic-d_3 acid-d (99.5 atom% d, from Sigma). The NMR sample tubes were then sealed under a blanket of nitrogen.

Fast atom bombardment (FAB) and desorption-chemical ionization (DCI) mass spectral analyses were performed on a VG Model 70G double-focusing mass spectrometer by Dr. Alvin Marcus of the Department of Chemistry at the University of Pittsburgh. FAB used 8 kV xenon atom bombardment, with glycerol as sample matrix. DCI was performed using isobutane as the reagent gas, at a resolution of 1500. The sample was in methanol: water (4:1).

Results

Identification of exotoxin peak. Since no authentic sample of β-exotoxin was available, it was first necessary to identify which peak in a chromatogram corresponded to this molecule. This was done by correlating the presence of a peak in the chromatogram with fly bioassay activity, and ultimately by isolating the peak and demonstrating that its properties were identical to those reported in the literature (13, 14, 17, 18, 21). Figure 2 compares the elution pattern detected at 260 nm, obtained from injecting an autoclaved C2 culture supernatant from dB1:1 (a strain identical to HD2-1, the variety *thuringiensis* type strain, which is known to produce exotoxin), with that obtained from a culture of HD1-1 (the exotoxin-negative variety *kurstaki* strain, used in many BT products). The major peak detected in dB1:1 that was missing in HD1-1 was tentatively identified as exotoxin, and is indicated by an arrow. Interestingly, another peak was present in HD1-1 which was missing from dB1:1. The presumptive exotoxin peak in dB1:1 was unchanged by autoclaving at pH 6.5, but disappeared upon autoclaving at pH 3, conditions expected to destroy exotoxin (18).

Cultures of HD-2 derivatives (22) were tested by HPLC, and all of the crystal-producing (Cry^+) variants produced exotoxin (as measured by fly bioassay), and showed the presumptive exotoxin peak on HPLC. Crystal-negative (Cry^-) variants, which were cured

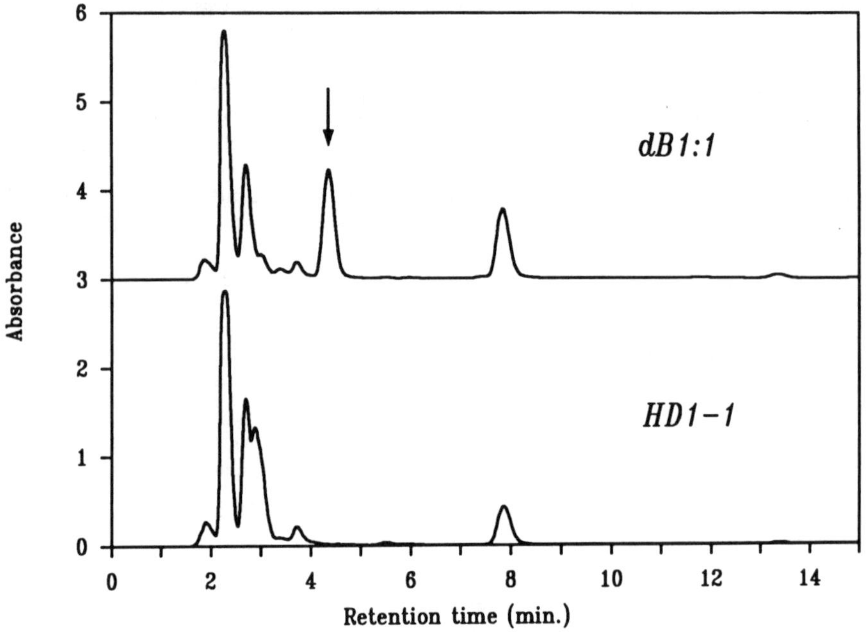

Figure 2. Identification of presumptive exotoxin peak on HPLC. (Reproduced with permission from Levinson, B.L., et al. J. Bacteriol., in press. Copyright 1990 ASM.)

of the 75 MDa crystal protein plasmid, did not produce exotoxin (Exo⁻). In addition, cultures of transconjugants, made by using the Cry⁻ strain HD73-26 as a recipient, were tested by HPLC, and only those which acquired the 75 MDa plasmid had the peak suspected to be exotoxin, and these were toxic to flies. Examples of the elution patterns obtained from a Cry^-Exo^- cured derivative (HD2-2) and a Cry^+Exo^+ transconjugant (HD73-26-19) are shown in Figure 3. Note that the transconjugant is an exotoxin producing subspecies *kurstaki* strain.

The peak presumed to be exotoxin was collected from HPLC separations of a concentrated C2 culture of dB1:1, neutralized and lyophilized. The material isolated from an overloaded analytical HPLC column appeared virtually homogeneous upon reinjection, and had properties identical to those for material further purified by gel filtration on Sephadex G-10. In order to confirm that this material was exotoxin, it was bioassayed, and found to be the only material eluting which had house fly toxicity. Exotoxin thus prepared had the expected absorption spectrum, with a maximum at 259 nm, minimum at 230 nm, and 258/238 = 2.5 (13, 21). It could be partially converted into dephosphorylated exotoxin by incubation with alkaline or acid phosphatase, as reported (14, 17). As expected for a less polar molecule, the dephosphorylated peak eluted significantly later from the column (18). Autoclaving at neutral pH had no effect on the exotoxin peak, while autoclaving at pH 3 gave a number of breakdown products. When kept at room temperature and pH 3, exotoxin is fairly stable over the time necessary for a set of separations; 13% dephosphorylation was observed in a sample reanalysed after one day. Even a neutral solution of exotoxin, kept at 4°C, slowly dephosphorylated, but could be used as a standard for several months. For quantitation, we used an absorbance of 18 for 1 mg/ml (13).

Comparison of fly and HPLC assay. In almost every case we considered (more than 30 strains), the two methods of determining exotoxin agree for the qualitative determination of exotoxin. HPLC allows quantitation, and we have found that the amount of exotoxin produced varies considerably from strain to strain. In a minimal medium, we have found that exotoxin production has varied from as little as 1 μg/ml to as much as 350 μg/ml, with the vast majority of the strains producing from 100-200 μg/ml (data not shown). We do not know whether the fly assay would detect toxicity in the strains producing very low levels of exotoxin in minimal medium, since all strains tested by fly assay produced at least 90 μg/ml.

The use of HPLC is somewhat more problematic when production samples or formulated materials are to be tested. Under the conditions described in Levinson, B.L., et al. (J. Bacteriol., in press), some medium components can be present in such high amounts that they start to overlap the elution time of exotoxin. For example, a sample of formulated product, known to be free of exotoxin, was spiked with 100 ppm of purified exotoxin, and both samples were run in this HPLC system. Figure 4 shows that this level of exotoxin can be detected by HPLC, but that 10 ppm would probably not be detected. This issue might be addressed by slight alteration of the running conditions. For instance, we have shown that the retention time is temperature-dependent, with the expected decrease in retention at elevated temperature (Figure 5).

When we tested BT strains (including the type strains) of 10 serotypes (1 *thuringiensis*, 3ab *kurstaki*, 4ac *kenyae*, 8ab *morrisoni*, 9 *tolworthi*, 10 *darmstadiensis*, 11ab *toumanoffi*, 17 *tohokuensis*, 18 *kumamotoensis*, and 20ab *yunnanensis*), the type strains of serotypes 1, 9 and 10 were positive for both the exotoxin peak on HPLC and for fly bioassay toxicity. Interestingly, although three naturally-occurring strains of variety *kurstaki*, including the type strain, were negative on HPLC and fly bioassay, a fourth *kurstaki* strain, NRD-12, was positive on both. There was one notable exception to the correlation of HPLC results to those from fly bioassay. HD-12 (8ab, *morrisoni*) was

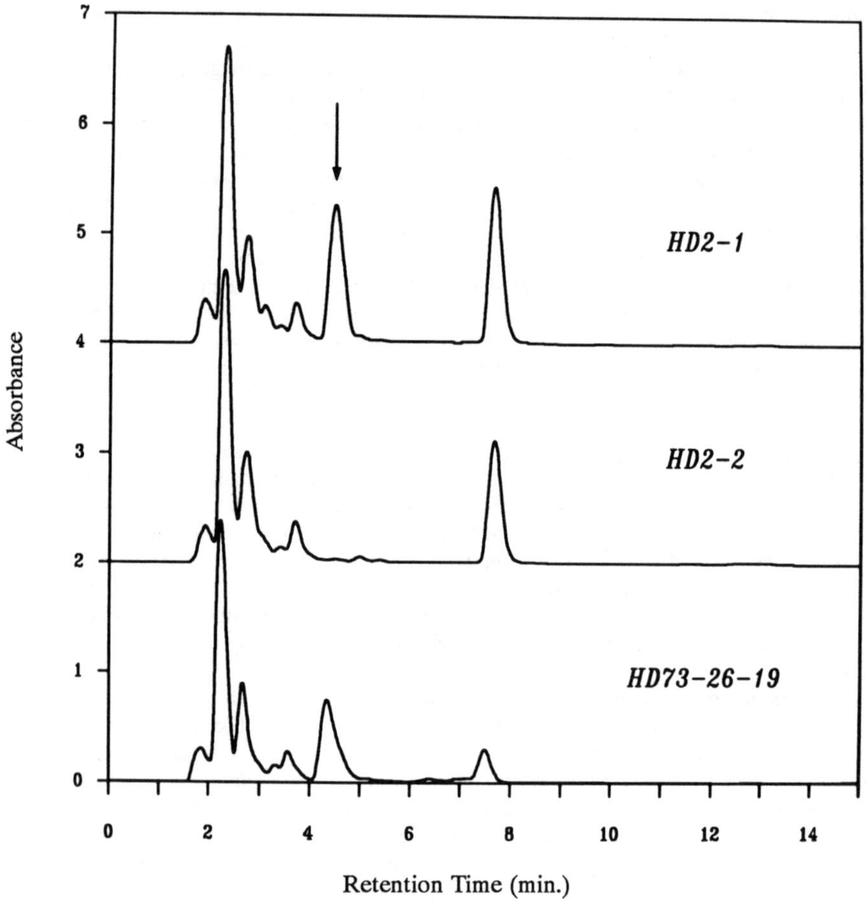

Figure 3. Exotoxin is plasmid-encoded in HD-2.

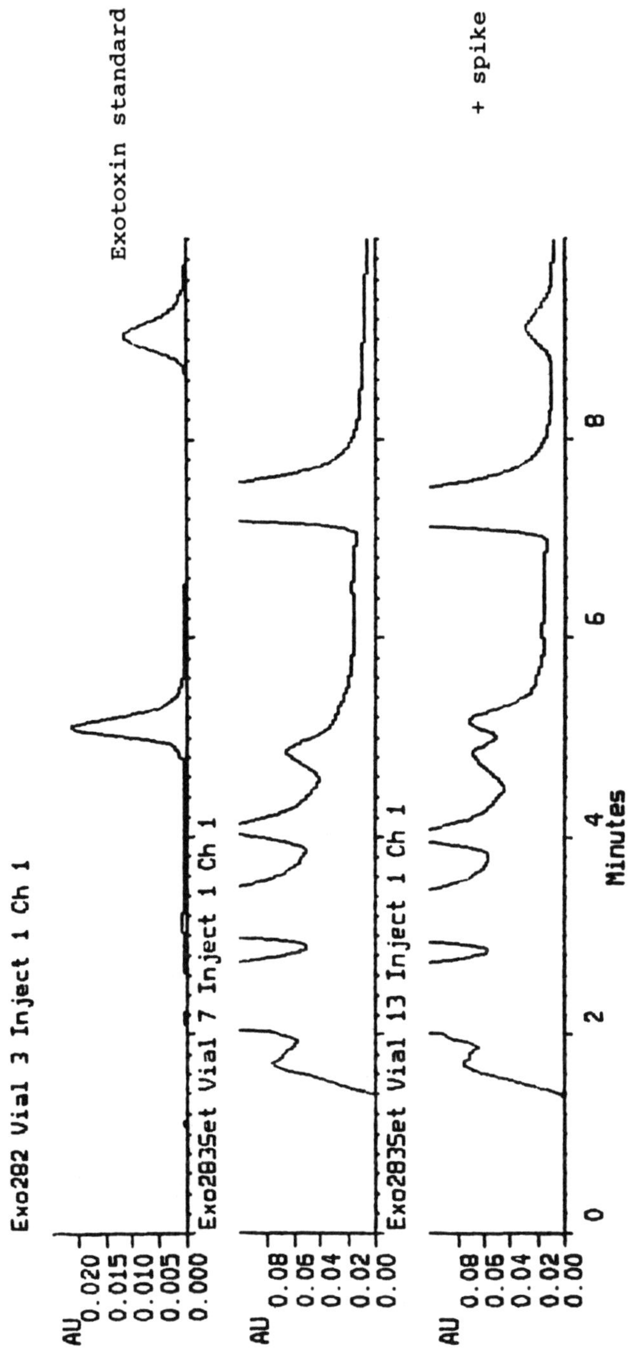

Figure 4. Test for exotoxin in formulated product, demonstrating a detection limit <100 ppm.

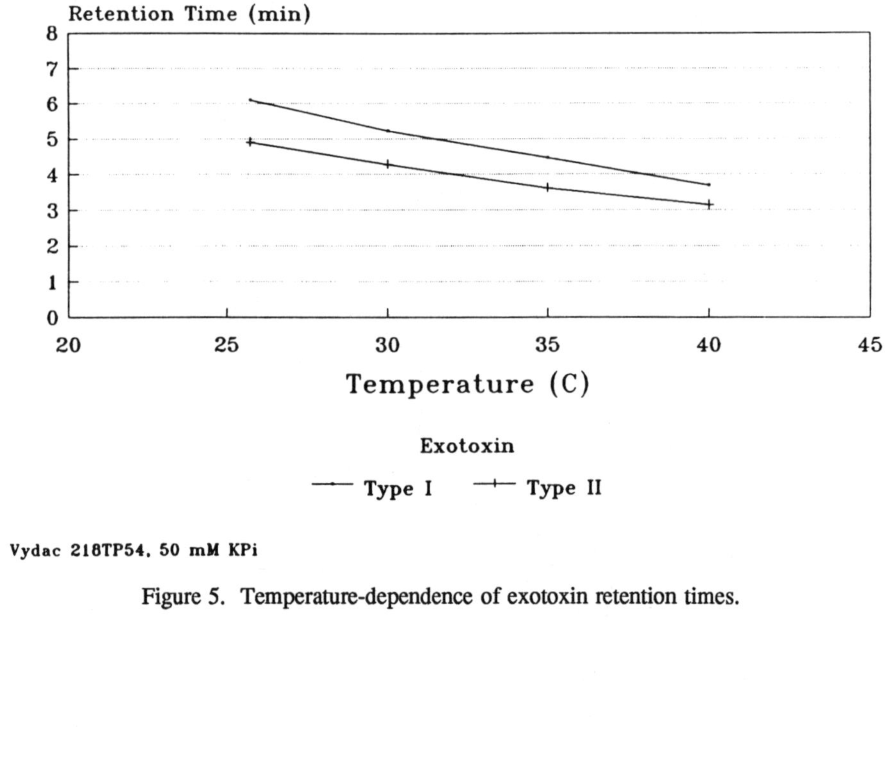

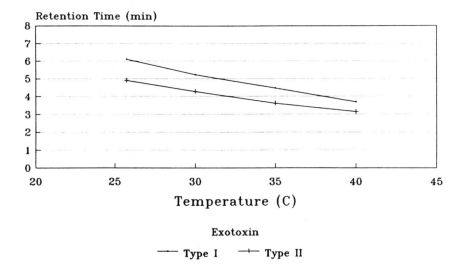

Figure 5. Temperature-dependence of exotoxin retention times.

negative by HPLC, but was exotoxin-positive by fly assay. This is the type strain for its serotype, and was previously reported as positive (23). This strain was studied in more detail (see below). In no case was a peak at the location of authentic exotoxin found for a strain which gave a negative fly assay result.

Correlation of exotoxin production with plasmids. The results above showed, by curing and by transfer into a Cry$^-$ host, that the 75 MDa plasmid from HD-2 encoded the fly-toxic activity. We examined other strains, and were able to assign exotoxin production to a plasmid in six strains of four serotypes. Examples of the data used to make these assignments are shown in Figure 6. Figure 6a displays an SDS-polyacrylamide gel, and Figure 6b shows an agarose gel of the same BT strains. In (a), the lane labelled "Standards" contained Bio-Rad Laboratories' (Richmond, CA) high and low molecular-weight standard proteins, with their masses (in kDa) given on the left margin. In (b), the standard lane shows the plasmids of strain HD1-1 (24) as size standards, with their masses (in MDa) given on the left margin. "Cry/Exo" indicates whether the strain produces δ-endotoxin crystals and exotoxin (+) or makes neither (-).

The wild-type, serotype-1 strain HD2-1 produces β-exotoxin and two distinct P1 proteins of 150 and 138 kDa. Loss of the 75 MDa plasmid generated strain HD2-2, which is Cry$^-$ (on most media), and fails to produce P1 protein or β-exotoxin. Transfer of the 75 MDa plasmid into the Exo$^-$Cry$^-$ strain HD73-26 generated the transconjugant HD73-26-18, which is Exo$^+$ and produces the larger, 150 kDa P1 protein. Transfer of both the 75 MDa and 54 MDa plasmids yielded the transconjugant HD73-26-19, which is Exo$^+$ and produces both the 150 and 138 kDa P1 proteins.

Similar explanations are made for the other strains shown. The wild-type, serotype-9 strain HD13-1 is Exo$^+$ and produces P1 proteins of at least two types, as well as P2 protein. Loss of the 44 MDa plasmid in HD13-3 eliminated most of the P1 protein production. Loss of both the 44 MDa and 110 MDa plasmids in HD13-2 eliminated β-exotoxin production, as well as all crystal protein production (P1 and P2). Loss of a single, large plasmid in the Exo$^+$P1$^+$ serotype-10 strains HD146-1 and HD498-1 gave the Exo$^-$P1$^-$ derivatives HD146-2 and HD498-2. The serotype-9 strain HD-537a produces β-exotoxin and several types of parasporal crystals. Partially-cured variants such as HD537c-75, which is Exo$^-$Cry$^-$, were used to assign the β-exotoxin and crystal protein genes unambiguously to a single large plasmid of 110 MDa.

The data for these strains are summarized in Table I, and demonstrate that a variety of plasmids, as defined by their size and by which endotoxin genes they carry, can encode exotoxin. In each case, parasporal inclusions of the type shown are found when the indicated plasmid is the only toxin plasmid present, and strains cured of this plasmid are Cry$^-$. In HD146-1, HD498-1 and HD537-1, this is the only crystal-encoding plasmid. The appropriately-sized proteins were found on polyacrylamide gels of culture samples only when crystals were observed (Figure 6). Thus, in all of these cases, the β-exotoxin plasmid is also a δ-endotoxin plasmid. In at least one strain (HD-2), this plasmid is transmissible by the natural conjugative system operating in *Bacillus thuringiensis*.

The case for the *kurstaki* strain NRD-12 is somewhat different. As can be seen from Table II, crystal production can be completely eliminated through the curing of the 115 MDa and 46 MDa plasmids, without loss of exotoxin production. This is the first well-defined case of a Cry$^-$ exotoxin-producing BT strain, although others have reported the existence of such strains (23, 25). Evidence for the involvement of the 77 MDa plasmid in exotoxin production in this strain comes from the loss of this plasmid, and concomitant appearance of a 55 MDa plasmid, in strain NRD12-10. It is presumed that this new plasmid arose by deletion from the 77 MDa plasmid of a fragment encoding exotoxin.

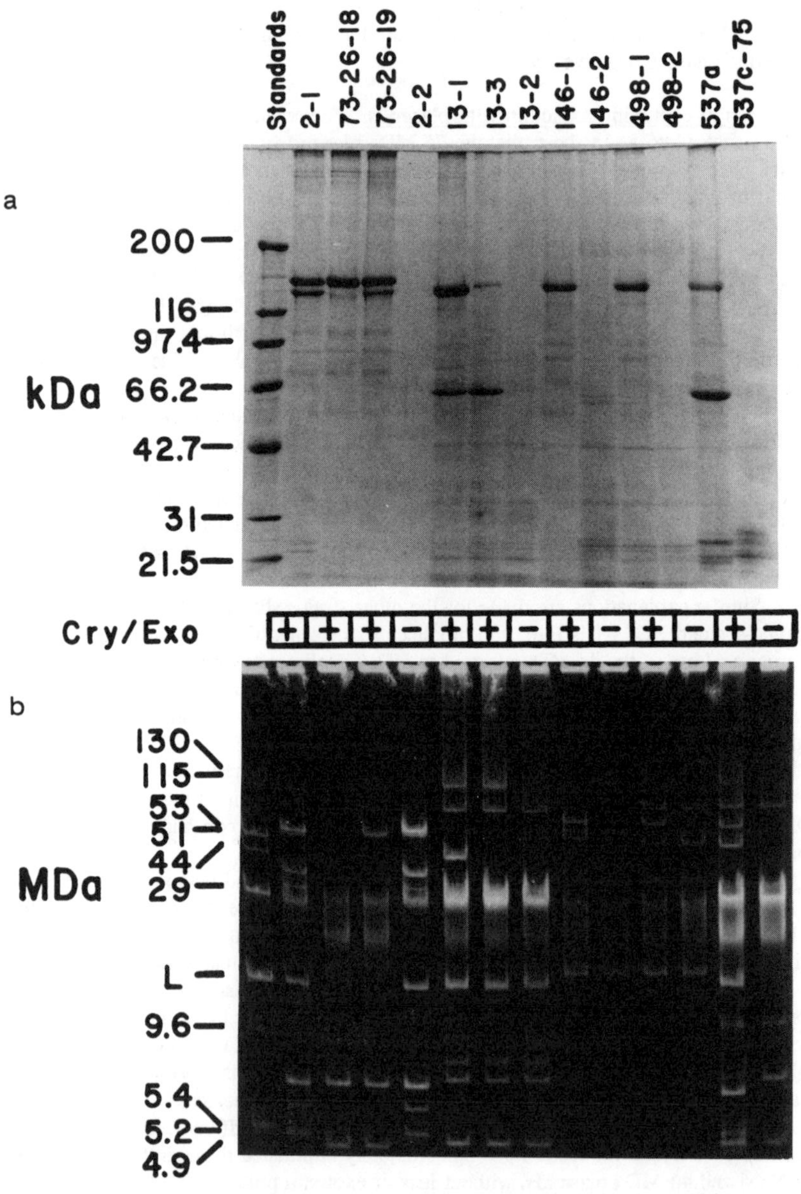

Figure 6. SDS-polyacrylamide (a) and agarose (b) gels of some exotoxin-producing BT strains, and some derivatives and transconjugants, demonstrating how exotoxin production is assigned to plasmids. (Reproduced with permission from Levinson, B.L., et al. J. Bacteriol., in press. Copyright 1990 ASM.)

TABLE I. Assignment of exotoxin production to plasmids in Bacillus thuringiensis

Strain	Serotype	Plasmid	Crystal†	Proteins*	Evidence**
HD-2	*thuringiensis*	75	P1	150	C,T
HD-13	*tolworthi*	110	P1,P2	140,67	C
HD-146	*darmstadiensis*	70	P1	140	C
HD-498	*darmstadiensis*	65	P1	140	C
HD-537	*tolworthi*	110	P1,P2,dot	135,67,23,21	C

† Morphology of delta-endotoxin crystal encoded by same plasmid.
* Approximate size (in kDa) on SDS gel of proteins encoded by same plasmid.
** C = curing; T = transfer into cry^- host.

TABLE II. Tentative identification of exotoxin plasmid in NRD-12 (*kurstaki*)

	Plasmid (MDa)										
Strain	130	115^+(P1,P2)	77	53	49	46^+([P1])	33	LDE	5.4	5.2	Phenotype
NRD-12	+	+	+	+	+	+	+	+	+	+	Cry^+Exo^+
NRD12-4	+	+	+	+	+	−	+	+	+	+	Cry^+Exo^+
NRD12-6	+	+	+	−	−	−	+	+	+	+	Cry^+Exo^+
NRD12-7	+	−	+	−	−	−	+	−	−	−	Cry^-Exo^+
NRD12-8	+	−	+	−	−	−	+	−	+	+	Cry^-Exo^+
NRD12-9	+	−	+	−	−	−	+	−	−	+	Cry^-Exo^+
NRD12-10	+	−	→55	−	−	−	+	+	+	+	Cry^-Exo^-

Note: All strains contain 9.6, 4.9 and 1.4 MDa plasmids.

Isolation and partial characterization of a new beta-exotoxin ("Type II"). Cultures of HD-12 did not exhibit the typical exotoxin peak on HPLC, but did have fly larvicidal activity. We isolated a fraction with the biological activity by HPLC of the concentrated supernatant from HD-12 (Figure 7). This material eluted earlier than authentic exotoxin, and hence is a different, less hydrophobic molecule, which we call Type II exotoxin (Levinson, B.L., et al., J. Bacteriol., in press).

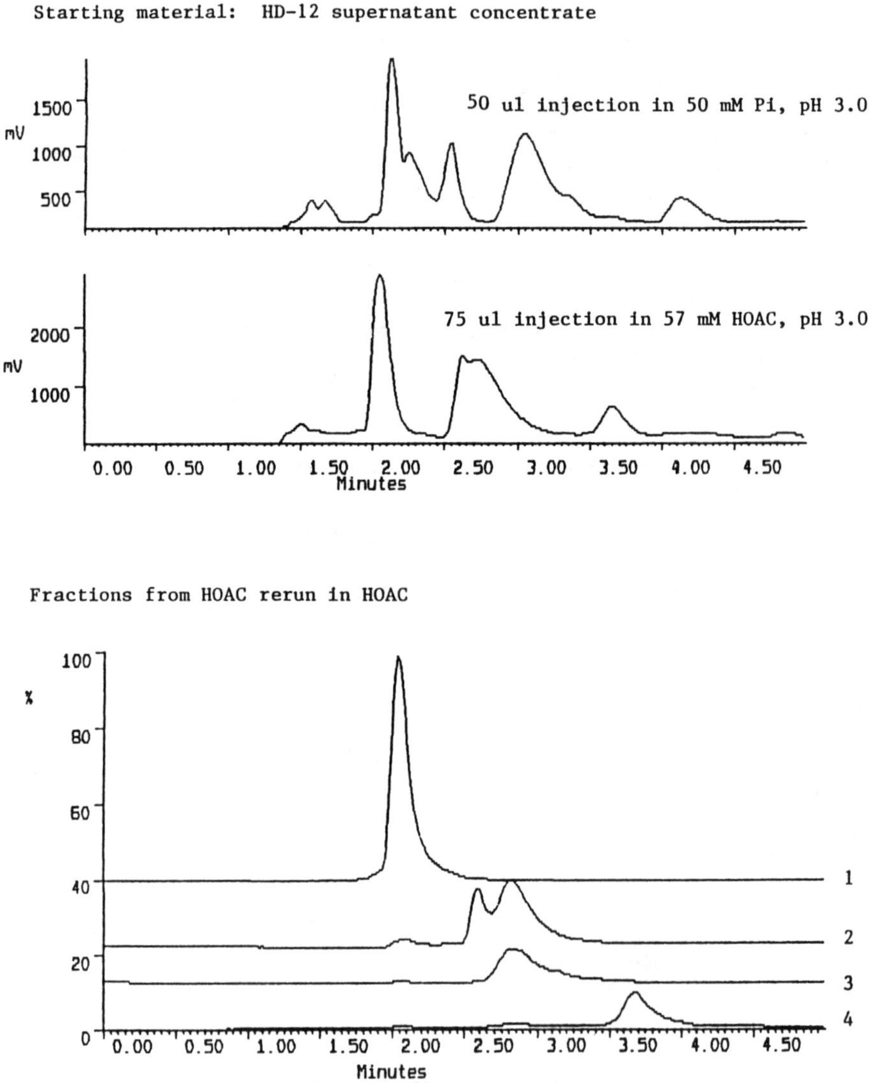

Figure 7. Isolation of a fraction from HD-12 containing exotoxin Type II.

Fractions from HOAC rerun in Pi

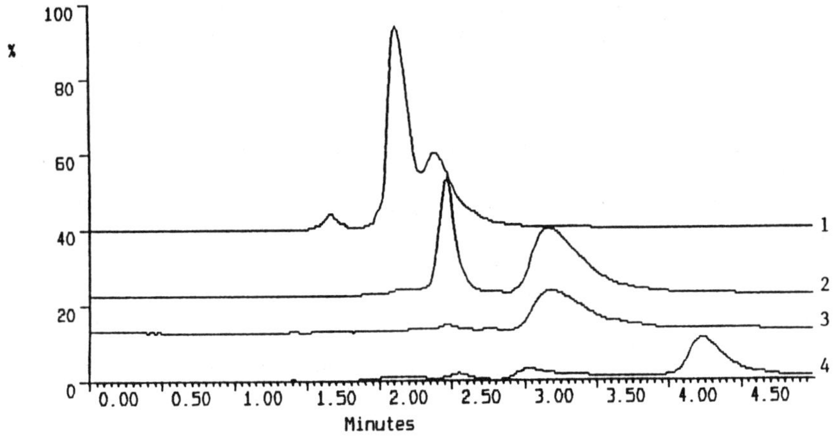

CHARACTERIZATION OF HPLC FRACTIONS

Fraction	HOAC ret. time	UV max	UV min	Fly mortality
1	2.06	254	230	0%
2	2.61 + 2.83	224	---	0%
3	2.84	249	229	0%
4	3.69	261	233	100%

Figure 7. Continued.

The major peak in the toxic fraction was dephosphorylated by either alkaline phosphatase or acid phosphatase, but to a different extent than Type I exotoxin. Figure 8 shows a comparison of the elution patterns of Types I and II exotoxin, before and after treatment with alkaline phosphatase. Identical samples of purified exotoxin were incubated in 100 mM $(NH_4)_2CO_3$, 0.125 mM $MgCl_2$, 0.2 mM $ZnCl_2$, with or without 1.275 mg/ml alkaline phosphatase (bovine intestinal mucosa, Sigma Type VII-SA) for 96 hours at 37°C, then prepared for HPLC. The arrows indicate the positions of the exotoxin peaks, while the asterisks indicate the positions of the dephosphorylated products. The reaction was run on a sample of lower purity than the material shown in Figure 7, but it is clear that Type I exotoxin is only partly dephosphorylated, while Type II has almost completely reacted.

The suspect peak was isolated in virtually pure form (by HPLC), and characterized by UV absorption. Figure 9 compares the UV spectra of Types I and II exotoxin, as a function of pH. Equal samples of purified exotoxin were diluted into 50 mM phosphate buffer at the indicated pH. The pH-dependent spectral shift in the Type II spectrum is reminiscent of that of the uracil ring. In uracil, this shift of absorbance with pH is due to lactam-lactim (keto-enol) isomerization. The absorption coefficient ($E^{1cm}_{1\%}$) was estimated to be 118 for Type II exotoxin, versus 180 for Type I, and this difference is also consistent with the uracil versus adenine chromophore. The dephosphorylation and UV spectral data suggested that Type II β-exotoxin might simply be the analogue of Type I exotoxin, containing some form of uracil, instead of adenine.

In order to test this hypothesis, further analyses were performed. ^{1}H-NMR spectra were run on samples of each exotoxin, at 0.5-1 mg/ml in deuterated DMSO. The down-field regions of these spectra are shown in Figure 10. In agreement with the literature (26), Type I β-exotoxin, shown in the top trace, exhibits the expected two adenine-ring single signals at 8.14 and 8.34 ppm, and the 5.9 ppm doublet due to the ribose 4′-proton. Type II β-exotoxin, shown in the bottom trace, does not have the diagnostic uracil ring doublets, expected at 5.9 and 7.9 ppm, but rather, a singlet at 7.9 ppm. This suggests that the uracil would have to be found as the pseudouridine moiety, in which attachment to the ribose ring is at the position which would otherwise be occupied by the proton expected to give a 5.9 ppm signal. That would account for the presence of the 7.9 ppm proton as a singlet as well. The pseudouridine attachment is found in some transfer RNAs. Other expected peaks are absent as well, notably the proton signal at 5.9 ppm due to the ribose 4′-proton. Thus, although uracil may be present, much of the structure must be significantly different from that of Type I exotoxin, and Type II exotoxin is not simply the uracil analogue.

Mass spectral analysis was also attempted, which gave mediocre signals by both methods tried. FAB emphasizes polar compounds, while DCI is better for non-polar materials. Molecules of intermediate polarity, such as nucleosides, could be detected by either method. Both methods gave 481 mass units as the largest peak, which is too small for molecule to be the analogue of Type I exotoxin, and not obviously obtained as a breakdown fragment thereof. None of the elemental compositions proposed from the FAB data is consistent with this concept either. It will require further work to elucidate the true structure of this interesting molecule.

Autoclaved HD-12 culture supernatant or a concentrate reconstituted from lyophilized broth ultrafiltrate, was tested versus HD-2 for activity against a variety of insects, and the bioassay data are summarized in Table III. Treatment with HD-2 supernatant (containing Type I exotoxin) at 1/10 dilution in a surface-contamination bioassay killed a high percentage of the *Spodoptera exigua*, *Trichoplusia ni*, *Heliothis virescens* or *Heliothis zea* larvae tested (all order *Lepidoptera*), but only 10% of *Diabrotica undecimpunctata howardi* (order *Coleoptera*). Only half the *Diabrotica* were killed by undiluted HD-2 culture, with the survivors severely stunted.

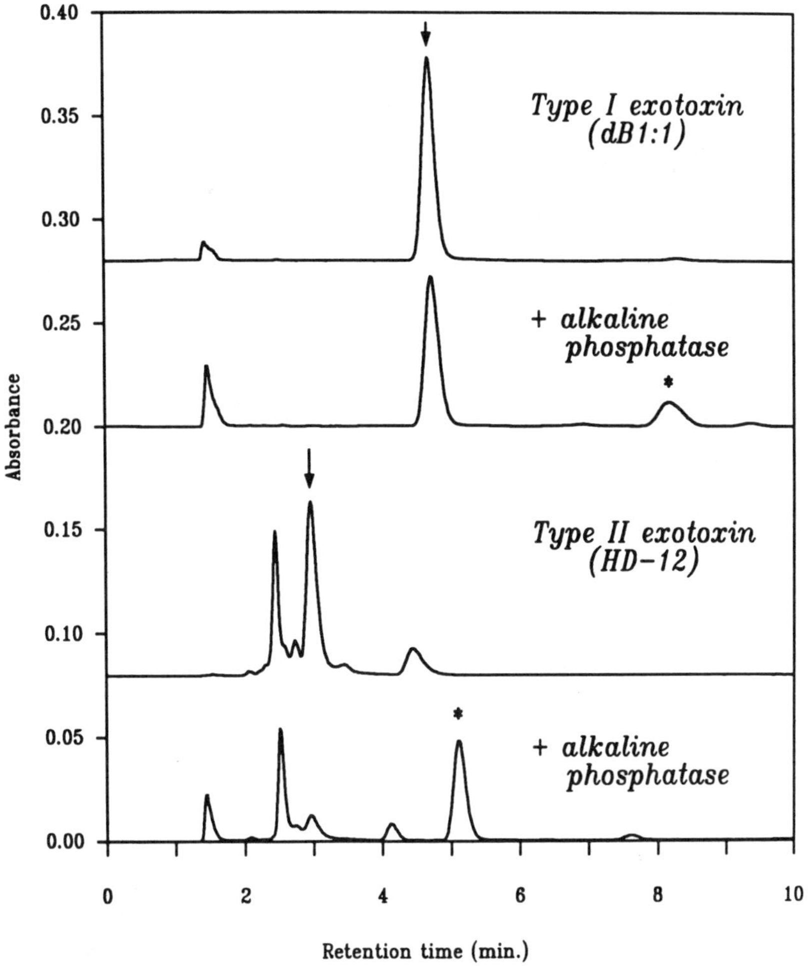

Figure 8. Comparison of Types I and II exotoxin, and their dephosphorylation by alkaline phosphatase. (Reproduced with permission from Levinson, B.L., et al. J. Bacteriol., in press. Copyright 1990 ASM.)

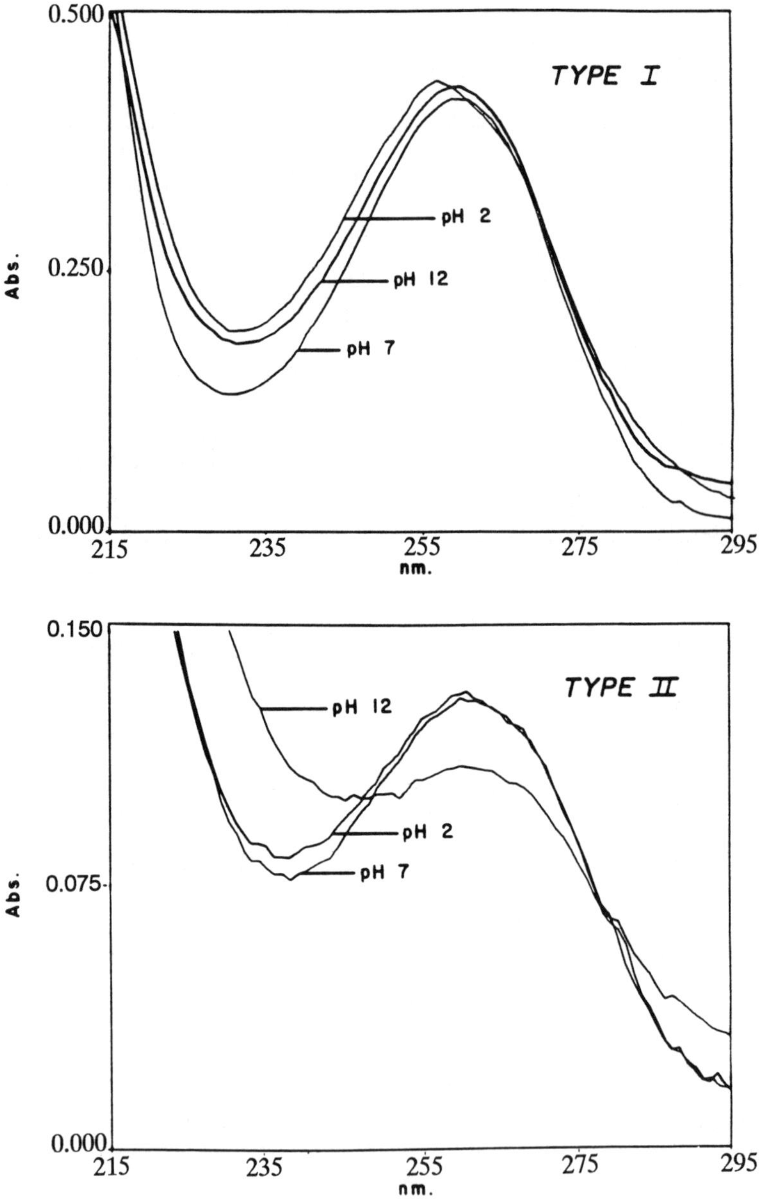

Figure 9. UV characterization of Types I and II exotoxin, as a function of pH. (Reproduced with permission from Levinson, B.L, et al. J. Bacteriol., in press. Copyright 1990 ASM.)

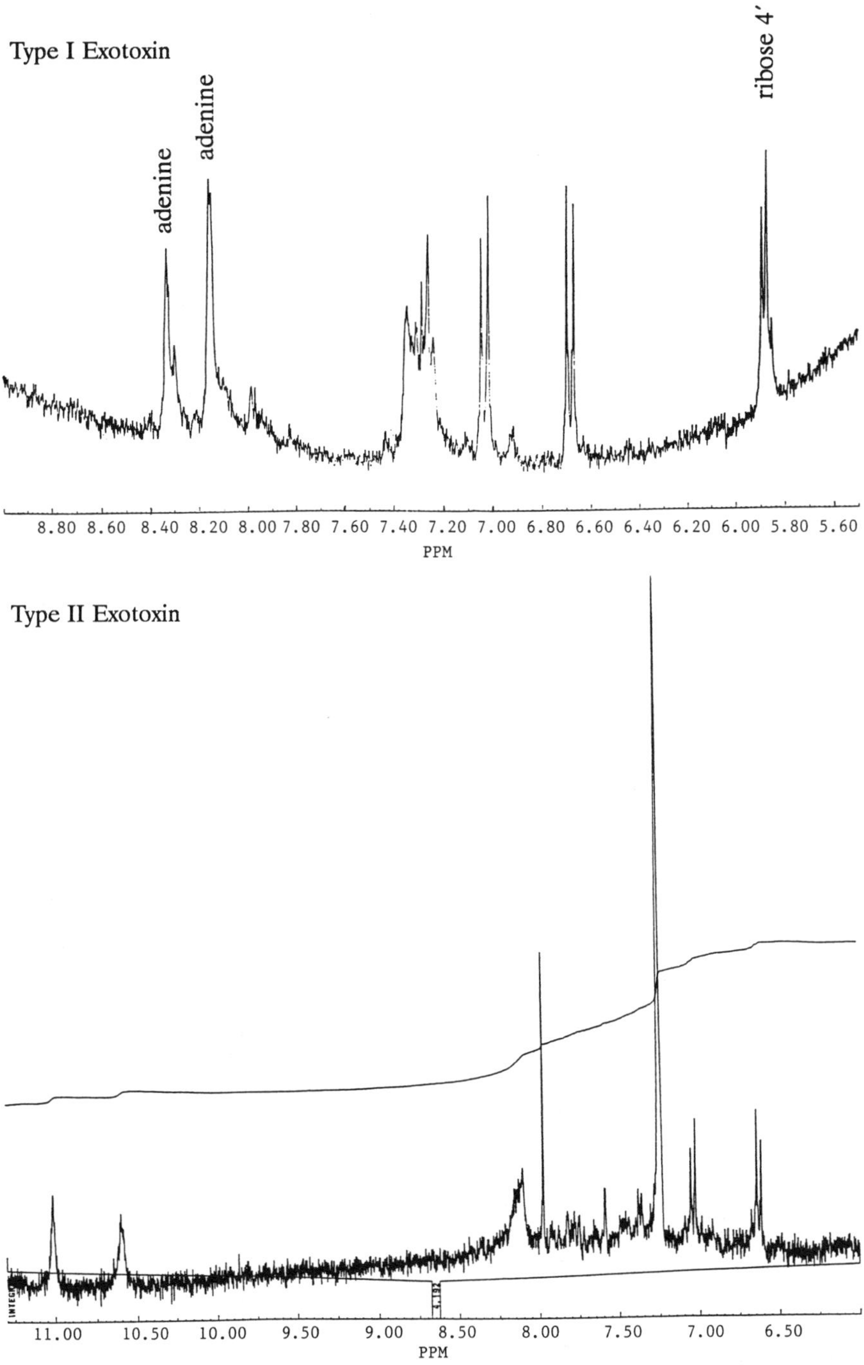

Figure 10. 300 MHz proton NMR spectra of Types I and II exotoxin, showing the downfield signals from the ring protons.

TABLE III. Bioassays of autoclaved broth supernatants (C2 medium)†

		% Mortality for (Insect)*					
Treatment		SE	TN	HV	HZ	SCRW	CPB
Control (water only)		0,0	3,10	3,43	0,7	0,20	10
HD2 (Type I):	1x	-	-	-	-	65,50,35	100
	0.5x	-	-	-	-	-	70
	0.25x	-	-	-	-	-	80
	0.10x	100,80	100,60	100,90	90,80	10,10	-
	0.03x	0	10	10	10	-	-
	0.01x	0	0	10	0	-	-
HD12 (Type II):	17x	-	-	-	-	100,94	-
	1.7x	100	100	100	100	45,25	-
	1x	-	-	-	-	0	100
	0.5x	50	100	80	10	-	90
	0.25x	-	-	-	-	-	60
	0.17x	20	70	20	10	-	-
	0.10x	20	10	10	20	10	-
	0.01x	0	0	0	0	-	-

† Concentrated materials were reconstituted from lyophilized broth ultrafiltrate.

* SE = *Spodoptera exigua*; TN = *Trichoplusia ni*; HV = *Heliothis virescens*; HZ = *Heliothis zea*; SCRW = *Diabrotica undecimpunctata howardi* (southern corn root worm); CPB = *Leptinotarsa decemlineata* (Colorado potato beetle). For SE, TN, HV and HZ, 30 insects were tested per treatment; for SCRW and CPB, 10 insects were tested. Dash = not tested. Mortality differences of 10-20% are not significant.

In contrast, HD-12 culture supernatant (containing Type II exotoxin) at 1/10 dilution killed only 10-20% of these insects. However, undiluted HD-2 and HD-12 supernatants both killed 100% of *Leptinotarsa decemlineata* (also order *Coleoptera*). The LC-50's of the two exotoxins for this insect, on the basis of culture volume, were comparable, being about 4-8 μl C2 supernatant/cup. At higher doses, however, Type II exotoxin was capable of killing all of the insects tested, including *Diabrotica.* Based on the measured absorption coefficients, and peak areas found on HPLC, we estimate that on a weight basis, Type II exotoxin is present in the HD-12 culture supernatant in only about 1/17 as much as the Type I exotoxin is in the HD-2 culture supernatant. Therefore, the toxicity of Type II exotoxin is actually greater, on a weight basis, for every insect tested. Furthermore, the specificity is different: *Coleoptera,* and especially *Leptinotarsa decemlineata,* are much more sensitive to Type II than Type I β-exotoxin. A summary of the comparative properties of Types I and II β-exotoxin is found in Table IV.

Discussion

We have been able to improve the separation of *Bacillus thuringiensis* β-exotoxin by HPLC (18), by using a wide-pore reverse phase column. This increased both resolution and speed of separation. The improved resolution allows categorical classification of *Bacillus thuringiensis* strains as to production of Type I β-exotoxin (found in the prototype exotoxin producer, HD-2 *thuringiensis*). There were no false positives; that is, all strains with a peak at the expected position have fly-toxic activity. We find this assay method simple, fast, and reliable for the exclusion from further development of Type I exotoxin-producing isolates. The reliable limit of detection (for the purposes of excluding strains for further development) for exotoxin by HPLC, under the conditions used for this study, and with a 50 μl injection, should be considered to be about 5 μg/ml in a culture supernatant. An otherwise interesting strain testing positive with a level lower than this should be repurified, in order to exclude the possibility that the exotoxin was produced by a minor (<1%) contamination with another, exotoxin-producing, bacterium. The detection level in HPLC could be pushed below the 1 μg/ml level with alternate detectors and microbore column techniques, but the problem soon becomes one of bacteriological purity rather than detection.

The high resolution of this HPLC method led to the discovery and isolation of a new toxin from strain HD-12 *morrisoni.* Some alteration of the separation conditions will be necessary in order to detect Type II exotoxin with the same reliability as Type I. Since, for example, the retention times of the two exotoxins change significantly with temperature, it may be possible to define conditions under which each of the exotoxins elutes independently of the other components. The existence of this newly-described Type II β-exotoxin raises another issue with respect to routine screening for exotoxins. It remains a formal possibility that other exotoxins exist. If so, they occur at a much lower frequency than Type I β-exotoxin, which we have found to be fairly common and widespread. Therefore, HPLC is valuable during the initial screening process, but it will remain necessary to look for exotoxin activity by bioassay before further strain development. Because exotoxin can be plasmid-encoded, it would also be wise to test for exotoxin before genetic material from a strain is transferred into a new host, either by conjugation or by genetic engineering techniques.

It is curious that exotoxin production seems limited to certain flagellar serotypes, even though mating can occur between varieties (27). In a way, variety has been used implicitly to predict the presence or absence of exotoxin production with some success (see e.g. reference 28). For example, there were no reports of exotoxin production in strains of variety kurstaki, until Ohba et al. reported the isolation of a single Exo$^+$ kurstaki strain (29).

TABLE IV. Comparison of the properties of the exotoxins from Bacillus thuringiensis

Both exotoxins are nucleoside phosphates, which are stable to autoclaving for 10 min at pH 7. Type II exotoxin is less stable to either alkaline (pH 10) or acid (pH 5) phosphatase-catalyzed dephosphorylation, while Type II is more stable at pH 3 in the absence of enzymes.

Other Characteristics	Type I	Type II
Relative partitioning on RP-HPLC:		
Capacity factor, k' (Vydac 218TP54):		
phosphorylated	2.4	1.1
dephosphorylated	4.9	2.6
Selectivity factor Type I/Type II:	phospho- = 2.2, dephospho- = 1.9	
Ultraviolet absorption:		
pH 7	max=259, min=231	max=261, min=235
pH 12	max=260, min=231	max=260 (reduced), min=245
abs. coeff. ($E_{1\%}^{1cm}$)	180	118
Molecular weight	701	<1000
largest fragment on mass spec.	684	481
NMR peaks (^{1}H)	downfield adenosine (8.14, 8.34 ppm) present	no adenosine, pseudouridine (?) (7.95 ppm) present, unidentified 7.21 ppm
Insecticidal properties: LC-50 for Leptinotarsa decemlineata	28 µg/cup	0.83 µg/cup

We have now found that another naturally-occurring strain of variety kurstaki (NRD-12) is Exo$^+$. We have also generated Exo$^+$ kurstaki strains (such as HD73-26-19) by conjugation. Although Exo$^+$ strains are still unknown in many serotypes, exotoxin production is nevertheless fairly broadly distributed among the serotypes. Clearly, serotyping alone is not sufficient to predict whether a strain may be an exotoxin producer. HPLC exotoxin determination can be a valuable time and labor saving tool, which can partly substitute for and confirm bioassay results. Ultimately, however, bioassay remains the true test of exotoxin production.

Acknowledgments

I thank K.J. Kasyan, S. Chiu, T.C. Currier, and J.M. González for work related to this project; T.B. Johnson, J.T. Greenplate and A.C. Slaney for insect bioassays; S.M. Brussock for providing C2 cultures in the initial part of this work; and C. Gawron-Burke for critical reading of the manuscript. L. Galya and A. Marcus were generous in their time and interest in the spectroscopic analysis of Type II exotoxin. I am indebted to M.E. Wagner Levinson for translations, editorial assistance, and patience.

Literature Cited

1. Lecadet, M. M.; H. de Barjac. In Pathogenesis of Invertebrate Microbial Diseases; E.W. Davidson, Ed.; Allanheld, Osmus & Co.: Totowa, NJ, 1981; pp 293-321.
2. Burgerjon, A. Annales de Parasitologie (Paris) 1974, 48, 835-844.
3. Šebesta, K., J. Farkaš, and K. Horská. In Microbial Control of Pests and Plant Diseases 1970-1980; H.D. Burges, Ed.; Academic Press, Inc.: New York, 1981; pp 249-281.
4. Vaňková, J. Folia Microbiol. 1978, 23, 162-174.
5. Horská, K., L. Kalvoda, and K. Šebesta. Collect. Czech. Chem. Commun. 1976, 41, 3837-3841.
6. Šebesta, K., and K. Horská. Biochim. Biophys. Acta 1970, 209, 357-367.
7. Šebesta, K., and H. Sternbach. FEBS Lett. 1970, 8, 233-235.
8. Smuckler, E. A., and A. A. Hadjiolov. Biochem. J. 1972, 129, 153-166.
9. Šebesta, K., K. Horská, and J. Vaňková. Collect. Czech. Chem. Commun. 1969, 34, 1786-1791.
10. Ignoffo, C.M., and I. Gard. J. Econ. Entomol. 1970, 63, 1987-1989.
11. Cantwell, G.E., A.M. Heimpel and M.J. Thompson. J. Insect Pathol. 1964, 6, 466-80.
12. Bond, R.P.M., C.B.C. Boyce, and S.J. French. Biochem. J. 1969, 114, 477-488.
13. Kim, Y.T., and H.T. Huang. J. Invert. Pathol. 1970, 15, 100-108.
14. Šebesta, K., K. Horská, and J. Vaňková. Coll. Czech. Chem. Commun. 1969, 34, 891-900.
15. Oehler, D.D., R.E. Gingrich, and M. Haufler. J. Agric. Food Chem. 1982, 30, 407-408.
16. Johnson, D.E., and R.E. Peterson. Eur. J. Appl. Microbiol. Biotechnol. 1983, 17, 231-234.
17. Bubenschikova, S.N., V.K. Kagramanova, and L.A. Baratova. J. Invert. Pathol. 1983, 42, 344-347.
18. Campbell, D.P., D.E. Dieball, and J.M. Brackett. J. Agric. Food Chem. 1987, 35, 156-158.
19. Höfte, H., and H.R. Whiteley. Microbiol. Rev. 1989, 53, 242-255.
20. Yamamoto, T., and T. Iizuka. Arch. Biochem. Biophys. 1983, 227, 223-241.

21. Benz, G. Experientia 1966, 22, 81-82.
22. González, J. M. Jr., H. T. Dulmage, and B. C. Carlton. Plasmid 1981, 5, 351-365.
23. de Barjac, H., A. Burgerjon, and A. Bonnefoi. J. Invert. Pathol. 1966, 8, 537-538.
24. Carlton, B.C., and J.M. González, Jr. In Mol. Biol. Microbial Differentiation; American Society for Microbiology: Washington, 1985; pp 246-252.
25. Shieh, T. R., and M. H. Rogoff. US Patent No. 3,758,383, 1973.
26. Farkaš, J., K. Šebesta, K. Horská, Z. Samek, L. Dolejš, and F. Šorm. Collect. Czech. Chem. Commun. 1977, 42, 909-929.
27. González, J.M. Jr., B.J. Brown, and B.C. Carlton. Proc. Natl. Acad. Sci. USA 1982, 79, 6951-6955.
28. Dulmage, H.T., and cooperators. In Microbial Control of Pests and Plant Diseases 1970-1980; H.D. Burges, Ed.; Academic Press, Inc.: New York, 1981; pp 193-222.
29. Ohba, M., A. Tantichodok, and K. Aizawa. J. Invert. Pathol. 1981, 38, 26-32.

RECEIVED April 9, 1990

INDEXES

Author Index

Affiliation Index

Subject Index

A

I

L

Q

R

U

V

W

Production: Donna Lucas
Indexing: Ann Maureen Rouhi
Acquisition: Cheryl Shanks

Elements typeset by Hot Type Ltd., Washington, DC
Printed and bound by Maple Press, York, PA

Paper meets minimum requirements of American National Standard for Information Sciences—Permanence of Paper for Printed Library Materials, ANSI Z39.48–1984 ∞

Other ACS Books

Chemical Structure Software for Personal Computers
Edited by Daniel E. Meyer, Wendy A. Warr, and Richard A. Love
ACS Professional Reference Book; 107 pp;
clothbound, ISBN 0–8412–1538–3; paperback, ISBN 0–8412–1539–1

Personal Computers for Scientists: A Byte at a Time
By Glenn I. Ouchi
276 pp; clothbound, ISBN 0–8412–1000–4; paperback, ISBN 0–8412–1001–2

Biotechnology and Materials Science: Chemistry for the Future
Edited by Mary L. Good
160 pp; clothbound, ISBN 0–8412–1472–7; paperback, ISBN 0–8412–1473–5

Polymeric Materials: Chemistry for the Future
By Joseph Alper and Gordon L. Nelson
110 pp; clothbound, ISBN 0–8412–1622–3; paperback, ISBN 0–8412–1613–4

The Language of Biotechnology: A Dictionary of Terms
By John M. Walker and Michael Cox
ACS Professional Reference Book; 256 pp;
clothbound, ISBN 0–8412–1489–1; paperback, ISBN 0–8412–1490–5

Cancer: The Outlaw Cell, Second Edition
Edited by Richard E. LaFond
274 pp; clothbound, ISBN 0–8412–1419–0; paperback, ISBN 0–8412–1420–4

Practical Statistics for the Physical Sciences
By Larry L. Havlicek
ACS Professional Reference Book; 198 pp; clothbound; ISBN 0–8412–1453–0

The Basics of Technical Communicating
By B. Edward Cain
ACS Professional Reference Book; 198 pp;
clothbound, ISBN 0–8412–1451–4; paperback, ISBN 0–8412–1452–2

The ACS Style Guide: A Manual for Authors and Editors
Edited by Janet S. Dodd
264 pp; clothbound, ISBN 0–8412–0917–0; paperback, ISBN 0–8412–0943–X

Chemistry and Crime: From Sherlock Holmes to Today's Courtroom
Edited by Samuel M. Gerber
135 pp; clothbound, ISBN 0–8412–0784–4; paperback, ISBN 0–8412–0785–2
